Ahead of Its Time

Ahead of Its Time:

THE ENGINEERING SOCIETIES LIBRARY, 1913-80

Ellis Mount

Linnet Books
1982

©1982 Ellis Mount
First published 1982 as a Linnet Book,
an imprint of The Shoe String Press, Inc.,
Hamden, Connecticut 06514

Printed in the United States of America

LIBRARY OF CONGRESS CATALOGING IN PUBLICATION DATA

Mount, Ellis.
    Ahead of its time.

    Includes bibliographical references and index.
    1. Engineering Societies Library—History.
I. Title.
Z675.E6M68   026.62   81-17155
ISBN 0-208-01913-8   AACR2

CONTENTS

ILLUSTRATIONS

TABLES

ABBREVIATIONS

AIChE	American Institute of Chemical Engineers
AIEE	American Institute of Electrical Engineers
AIME	American Institute of Mining, Metallurgical and Petroleum Engineers (also: American Institute of Mining Engineers; American Institute of Mining and Metallurgical Engineers)
ALA	American Library Association
ASCE	American Society of Civil Engineers
ASME	American Society of Mechanical Engineers
COMPENDEX	Computerized *Engineering Index*
EF	Engineering Foundation, Inc.
EI	*Engineering Index*
EJC	Engineers Joint Council
ESL	Engineering Societies Library
IEEE	Institute of Electrical and Electronics Engineers
IRE	Institute of Radio Engineers

NYSILL	New York State Interlibrary Loan program
SLA	Special Libraries Association
UEC	United Engineering Center
UES	United Engineering Society
UET	United Engineering Trustees, Inc.

Preface

Across the street from the attractive United Nations property on Manhattan's East Side is a modern multistoried building, the United Engineering Center. It houses the headquarters of several of the most important engineering societies in the United States, and on its second floor is the Engineering Societies Library, one of the leading technical libraries of the world. The purpose of this book is to describe the technical resources of this library and the significant contribution it has made to the furtherance of the goals of the engineering profession.

This book is based in part on my doctoral dissertation, which was written as a history of the Engineering Societies Library and covered its development in considerable detail.[1] This account—which emphasizes the many services offered by the Library, its networking activities, and the richness of its collection—updates the dissertation and thus spans the period from the creation of the Library in 1913 through 1980.

The first chapter presents a brief summary of the important features of the Library. Following that is a review of the nature and significance of engineering. Subsequent chapters cover the development of technical libraries, the creation of the Engineering Societies Library, and notable people associated with the Library. After that there are five chapters which discuss the Library's collection, cataloging activities, user services, involvement with networking and other cooperative activities, and relationship to *Engineering Index*. The final chapter considers the Library's future role.

Appreciation is expressed for the excellent cooperation given me by several members of the staff of the Engineering Societies Library, particularly its director, Kirk Cabeen; and for permission to reproduce illustrations and tables. The United Engineering Trustees were kind enough to grant me permission to reproduce other illustrations.

February 1981

1

A Brief Glimpse of The Library

One of the strengths of any society is its ready access to reliable, adequate technical data and also to information specialists able to aid in the collection, organization, and retrieval of such data. This combination of technical information and trained personnel able to work with the data constitutes technical resources, which are indispensable to the growth of science and engineering.

While there are many libraries and information centers which collectively serve a large number of scientists and engineers, there is in this country a remarkable library which was established for the primary purpose of serving engineers. Its collection is said to be one of the largest in the free world devoted exclusively to engineering. It was created when four relatively small libraries, which had been organized in the latter part of the nineteenth century by societies devoted to mechanical, mining, electrical, and civil engineering, were merged into one library serving all of these major engineering societies. Besides the charge to cater to the needs of members of these organizations, the Engineering Societies Library was also to function as a public library, thus aiding engineers not affiliated with its Founder Societies as well as the general public, including engineering students, businessmen and women, patent lawyers, and those in a number of other categories interested in engineering.

Since its creation in 1913 the Engineering Societies Library has grown to a place of importance in the world of technical information, internationally recognized as a most valuable asset

for engineers. This book aims at making its contributions more
widely known.

The Collection

For the benefit of readers not acquainted with the collection
of the Engineering Societies Library, a brief description is in
order. The collection concentrates on engineering in all its many
aspects; it does not include the pure sciences or the social sciences.
Although its total volume count is not so large as that of some
other notable technical libraries, its concentration on engineering
makes its strength in that subject evident. Other large technical
libraries may include technical reports and/or patents in their
collection count, while some collect in the fields of the pure
sciences, or medicine, or certain social sciences. These differences
make it difficult to compare collections of different libraries
solely on the basis of the total number of items.

The Library's collection includes a nearly complete set of un-
published conference papers sponsored by the engineering societies
affiliated with the library; no other library in the world can come
close to equalling this unique asset. Still another area of strength
is its holdings of pre-1830 engineering books: a study revealed
that the Library held nearly half of the material available from
this period, an enviable record few, if any, other libraries could
equal.[1] As for modern monographs, the Library contains all those
which are pertinent for a coverage of every aspect of engineering.

After the Library of the Engineering Societies (its original
name) was established in 1913, an effort was begun to consolidate
the collections of the three small libraries which were merged to
create it. These libraries represented societies of mining, mechanical, and electrical engineers, and their total stength was approximately 54,000 volumes, with 800 serials being received.[2] By the
close of 1915 some 16,000 duplicate volumes (mostly bound
periodicals) had been identified and removed.[3]

A sharp increase in the collection size occurred in 1917 when
the last of the four Founder Societies, the American Society of
Civil Engineers, voted to join its fellow societies in the building

housing them and to merge its library with what was soon to be officially named the Engineering Societies Library. This move brought an additional 67,000 volumes to the collection.[4] As shown in Appendix I the collection grew steadily, reaching the 200,000 mark in the 1950s, and a total of more than 267,000 volumes and maps were achieved by the end of 1980. By the latter date the Library was receiving more than 7,700 serials, including virtually all worthwhile journals and conference proceedings related to engineering. It had always had strong holdings of periodicals, representing all corners of the globe. These serials are of great importance, as is the case with any technical library, since they provide an excellent source of current, detailed scientific and technical information.

READER SERVICES[5]

Because of the dual set of clients which the Library was charged to serve in its charter (members of the sponsoring societies as well as the general public), its reading room accommodated only a fraction of its actual and potential users. While no known study shows the relationship of users to membership in the several engineering societies sponsoring the Library, the fact that the sponsoring societies include members located in all parts of the United States and many foreign countries has assured a broad geographical coverage for the Library's clientele. For decades a large part of its use has been through mail and the telephone as compared to personal visits. The fiscal year 1951/52 marked the first time the annual number of library users who did not appear in person—that is, the mail and telephone users—exceeded the number of those who did personally use the Library. By the end of 1980 the number of mail and telephone users constituted more than two-thirds of the total annual number of users. Thus, as time went on, for whatever reason, fewer users came in person to the Library relative to the number writing their requests or using the telephone; no doubt a large part of this development is due to the fact that local usage has gradually been outdistanced by use from outside the New York region. A survey of those who used the

Library's photocopy service in 1978 shows that only 24.6 percent were located in New York, New Jersey, and Pennsylvania. The remainder of the users' locations ranged from Delaware to Oregon, and many foreign countries. There is every indication that this same pattern of user locations held for many years before the study was made. Appendix II shows the total number of users as well as the number of telephone users over the years.

Early in its history the Library made provision for special service to those in a position to pay a fee for special searches or for translations. Photocopying, now one of the main sources of income for the Library, has been available since 1917. By the end of 1980 more than 600,000 copies were being made each year for customers, plus nearly 165,000 copies made on self-service machines. In 1967 the Library signed a contract with the New York State Library to serve as one of the backup libraries filling interlibrary loan requests that public libraries and the state library could not fill. Upon the inauguration of this program the Library began receiving daily teletyped requests from the state library, providing this service on a fee basis. The Library has achieved an enviable record for quick responses and a high percentage of items located for loaning. By 1980, after thirteen years of the contract, some $265,000 had been added to the Library's coffers from this service. Approximately four thousand requests are filled per year. Recently another service has been inaugurated which extends even further the value of the Library to users on an international basis. In 1980 it became one of the libraries listed as a source for items cited in the computerized database version of *Engineering Index,* as made available to thousands of librarians and engineers who are customers of the Lockheed Information Service. By electronic means searchers, working at their individual computer terminals, can order copies of documents and periodical articles cited by the databases, to be mailed from the Engineering Societies Library without further effort on the part of requestors, thus saving themselves time and money in obtaining needed materials.

Loans of books to members of sponsoring societies grew in number over the years, so that nearly 3,700 were being loaned each year by 1980, going to requestors in all sections of the United

States and Canada. The conducting of searches, both manual and computer-made, constituted a major activity for the reference librarians; the advent of the Library's own terminal in the 1970s brought new capabilities to this service due to the speed and efficiency possible with computerized searches. These searches were for many years an important part of the reader services of the Library; adding the capability for making computerized searches has kept the library abreast of the times and will help ensure its continued value to its many users.

BIBLIOGRAPHIC CONTROL OF THE COLLECTION

The Engineering Societies Library is one of the few major libraries in this country which maintains a classed catalog. It is proud of that distinction, and its catalog department has taken great care to keep the classification scheme, based on the Universal Decimal Classification, revised and up-to-date. It began on a high plane of quality when Margaret Mann, internationally known for her work as an excellent cataloger, established the present system in the 1920s.

There are other ways in which the collection can be searched, one of which was the filming of the classed subject catalog by the G. K. Hall Company—a set of thirteen volumes subsequently was published in 1963, with annual supplements issued until 1975. After that time entries of the Library's new books and conference proceedings have been added to two monthly indexes issued by G. K. Hall, its *Technology Book Guide* and its *Conference Publications Guide*. A completely new edition of the G. K. Hall catalog is scheduled for publication late in 1981.

Another means of ascertaining the holdings of the Library is through the internationally known *Engineering Index*, published by a separate but closely related organization located in the United Engineering Center, home of the Library. Aside from a small number of abstracts of technical reports furnished to the *Index* by a few government agencies, the bulk of the contents of each issue of the *Index* (probably as much as 98 percent) consists of abstracts of books, journals, and similar material, all of which

may be found in the Library. Thus virtually all of its periodical
articles and some of its books may be located by author or by
subject through *Engineering Index*, said to be the world's foremost
periodical index for engineers. There is probably no other technical
library in the world having such a widespread audience fully
aware of the details of its contents. There have long been close
ties between the two organizations, continuing over the years
which saw the *Index* grow stronger and more innovative, as with
the creation of COMPENDEX, its computerized version, in 1969.
The contract with the Lockheed Information Service, previously
mentioned, makes the utilization of the Library's collection
even simpler for users of COMPENDEX.

SUMMARY

The growth of the Engineering Societies Library, not just in
the size of its collection, but in its ability to aid engineers on an
international basis, is the significant point of this brief intro-
ductory chapter. The Library developed from an unwieldy confederation of four small libraries into a respected, well-used
organization serving major engineering societies as well as those
in the public sector desiring technical data.

The Library is now supported by eight sponsoring societies;
it should be noted that their combined membership in 1980 is
almost 420,000. The sponsors are: American Society of Civil
Engineers; American Institute of Mining, Metallurgical and
Petroleum Engineers; American Society of Mechanical Engineers;
Institute of Electrical and Electronics Engineers; American Insti-
tute of Chemical Engineers; American Society of Heating,
Refrigerating and Air-Conditioning Engineers; Illuminating
Engineering Society; and Society of Women Engineers.

2

The Significance And Development of Engineering

The Engineering Societies Library was founded by societies of engineers, a major source of its income is a portion of the dues paid by engineers to certain societies, and its main body of users is made up of engineers (or of information-gathering organizations serving engineers). Since its creation and its development are inextricably related to the nature of engineering and its practitioners, it seems appropriate to devote a chapter to an abbreviated analysis of the history and development of the engineering profession. This chapter will concentrate on (but not be restricted to) the state of affairs near the close of the nineteenth century, the era in which stirrings within the profession led to the formation of the Engineering Societies Library.

THE ROLE OF ENGINEERS IN SOCIETY

Engineers have been part of society almost from its earliest days. They may have been referred to in early eras as architects, since so much of their work was concerned with the design of structures and the supervision of their erection. Some may have been known simply as inventors or craftsmen. The term *ingenium* originated in the third century. It referred to certain military machines, and from this came the term *ingeniator*, meaning a person who could design such devices. It was not until the time of the Industrial Revolution that engineers' nonmilitary projects began to gain recognition. At that time a distinction began to be made between the military engineer and the civil engineer, the latter term pertaining to all those engaged in engineering projects of a civilian nature.[1]

But by whatever name engineers were known over the years, their contribution to society have made an indelible imprint. In many instances their work withstood the test of time better than the social institutions for which the work was done. As De Camp has so vividly put it:

> But through all the ages of history, one human institution — technology — has plodded ahead. While empires rose and fell, forms of government went through their erratic cycles, science flared up and guttered out, men burned each other over differences of creed, and the masses pursued bizarre fads and fashions, the engineers went ahead with raising their city walls, erecting their temples and palaces, paving their roads, digging their canals, tinkering with their machines, and soberly and rationally building upon the discoveries of those who had gone on before.[2]

While not every one would take such an exalted view of the work of engineers, there seems little argument that they have made countless long-term contributions to civilization over the years. In some cases the skill of ancient engineers defies belief in view of the relatively crude state of society and its tools at the time. For example, work was so accurate in constructing the Great Pyramid at Gizeh that each of its four sides, nearly eight hundred feet long, varied in length from each other by only an inch or so. Early wheeled chariots were built by the Sumerians in Babylonia probably as early as the twenty-fifth or twenty-sixth centuries B.C., a proof of the inventiveness of man at a primitive period.[3] Dozens of books could be cited which describe the lengthy list of the works of engineers over the centuries; many of these devices and projects changed the entire nature of the civilizations in which they came to light, and influenced succeeding generations.

As time went on, the nature of engineering gradually changed, one difference being the emergence of engineers who worked on projects on their own or who had civilian sponsors, no longer restricted to the royal patrons who had monopolized the time and energies of engineers for centuries. Another change evolved over many years in regard to the nature of engineering projects. For a very long time engineers were more or less restricted to work that we now call civil engineering—for example, the design of buildings or tunnels or highways as well as the supervision of the workers involved in such projects. But the growing need for machines or laborsaving devices led to the rise of mechanical engineers, who designed such objects as steam or internal combustion engines. The discovery of electricity led to the emergence of electrical engineers, those who could design communication equipment or provide society with electric generators. The need for mass production of certain industrial materials, such as acids or metals, gave birth to specialists who could design the equipment and processes necessary for manufacturing the products, these experts to be known as chemical engineers and metallurgical engineers. Industrial engineers appeared in technical circles performing such services as designing a factory in order to achieve the greatest efficiency, or to eliminate waste motions in the operations of complicated machinery. Other types of engineers came into being when advancements in technical knowledge created entirely new areas of specialization, as typified in more recent years by aeronautical engineers or nuclear engineers.

Discoveries in the so-called pure sciences—such as physics, chemistry, biology, geology and other earth sciences, or mathematics, to name some of the major scientific disciplines—were closely watched by engineers because such scientific advances often led engineers to the development of new processes, new products, or entirely new industries, or even the emergence of new engineering specialties. By 1885 a study by Lewis Haupt listed twenty categories into which he could divide the thousands of engineers named in his directory. The specialties he listed included mechanical, mining, gas, bridge, army, consulting, marine, sanitary, electrical, hydraulic, and chemical engineering.[4] The fact that there were thousands of engineers at that date was closely related to the great spurt in the development of new devices in the

United States since the Civil War. As Armytage points out in his detailed history of engineering, "Some index of the acceleration of American engineering can be obtained from the fact that, from its establishment in 1790 to 1860, the U.S. Patent Office granted 36,000 patents, but from 1860 to 1890 it granted no less than 440,000: more than a twelvefold increase in less than half the time."[5] The availability of these newly patented devices and processes was a spur to American industry. Likewise, industrialists began to support research and development staffs, which brought forth a multitude of patents.

While American ingenuity had been evidenced in the efforts of many people who were essentially tinkerers in their own small workshops, such inventors, as they could also be called, were gradually superseded by those with more formal schooling and with commercial sponsors. The need for expensive tools and long periods of development made it almost a necessity for creators of new products to have dependable financial backing that only a company or an institutional employer could provide. The growing cost of development of new engineering advances gradually led to this disappearance of the legendary lone inventor. There were, of course, exceptions to this; among the most notably successful was Thomas A. Edison, who, notwithstanding his limited formal schooling, produced numerous inventions that made lasting changes in our society. All his work was carried out in his own private laboratories.

Along with the changes in American society that led to increased concentration of wealth and people in cities came the ills of urban life. Engineering developments such as cheap electric power or fast mass transit systems could be said to have contributed not only to the creation of more efficient factories and concentrations of large apartment buildings, but also to polluted atmospheres and urban slums. When one reviews the development of engineering, one might ask whether or not engineers have consistently had sufficient concern about the effects of their work on their fellow human beings. On the other hand, many observers point out that top executives, not engineers, usually make final decisions about the social aspects of engineering projects.

Compared to conditions in the 1890s, the past few decades have brought increased governmental regulation of factors that affect the health and safety of our society, and also a greater tendency to enlist the aid of engineers to help solve some of the problems of the day. For example, more and more governmental units or private agencies have been making use of the talents of engineers in the planning and development stages of urban renewal projects. A paper written in 1973 by Eugene S. Fubini, a business executive, pointed out that engineers were becoming more aware of the social consequences of their work and were using interdisciplinary approaches to the technical problems of the urban areas. He felt that putting a man on the moon was a relatively straightforward matter in contrast with trying to cope with the decay of inner cities.[6]

A different era has arrived, in which engineers and their employers must plan their projects with a careful eye on the effects of their work on society. This was the viewpoint of a writer in the 1970s who said, "In order to address the needs of the contemporary world, the technical community will no longer be able to look at technical problems separate from the complex of social and economic issues in which they are embedded."[7] In recent years the role of engineers as full partners in the mammoth task of solving the problems of modern society—rural and urban—has been more firmly established. Engineers are not avoiding such an involvement; for some time the technical literature has been full of articles, books, and reports dealing with the social aspects of engineering projects and the part engineers are playing in such endeavors.

For example, a periodical article by Frey which surveyed the outlook for engineering stated: "Engineers are not only affected by changes within the profession, but by demands on engineering from without. There is great concern with the impact of engineering on the environment...." The engineer, the author went on to say, "is being asked to assume a collective sense of responsibility, but finds that the organizations he has developed to serve find it difficult to assume this role." Then he pointed out ways in which the engineer could best fulfill his responsibility both to his employer as well as to society as a whole.[8]

NUMBER AND TYPES OF ENGINEERS

Over the centuries since engineers were first identified there has been an irregular growth pattern in the number of engineers. While data for much of this period are scattered and uncoordinated, a reasonable picture of the development of the profession of engineers in the United States can be drawn for 1900-1970.

TABLE 1
GROWTH OF ENGINEERING SPECIALTIES[9]

TYPES OF ENGINEERS	NUMBER OF ENGINEERS (IN THOUSANDS)							
	1900	1910	1920	1930	1940	1950	1960	1970
Civil	20	40	56	88	97	128	158	175
Metallurgical & mining	3	7	11	14	12	23	31	21
Mechanical	14*	15	39	58	97	207	160	181
Electrical		15	27	58	65	110	185	286
Total (including types not listed)	38	77	134	218	297	543	872	1231

*Mechanical and electrical engineers for 1900 are counted together.

As the statistics show, different specializations changed their relative positions in the totals through the years. For example, electrical engineers were the same in rank as mechanical engineers in 1910 (15,000), but by 1970 there were over 100,000 more electrical engineers than mechanical engineers. All types of engineering specialties made impressive gains from 1940 to 1950, particularly mechanical engineers, then slowed down their growth thereafter, with the exception of electrical engineers. One other point to note is the complete domination of the engineering field for several decades by the engineers represented by the four original sponsoring societies (civil, mechanical, electrical, and mining) of the Engineering Societies Library. Later on other types reached significant size.

ENGINEERING EDUCATION

One of the generally accepted requirements of a profession is the establishment of formal courses of instruction, based on principles of major importance, with recognized goals for its students. Engineering, like medicine and law, did not at once possess a full-fledged educational program. In fact, when one considers the existence of engineers in society for thousands of years, it is significant that only for the past hundred years has there been anything like formal education at a professional level for engineers.

For centuries engineering education could not have been much different from the apprentice system so commonly used for the teaching of trades. While in certain periods in which science made relatively rapid strides there were formal technical schools, such as those in ancient Greece in the time of Plato (429-347 B.C.), most of the time there was only informal education for engineers. After the invention of printing the process of education was enormously aided by the publishing of texts, exemplified in 1556 by the *De Re Metallica*, a famous work by the German physician and mineralogist Agricola (the Latin form of his German name, Georg Bauer). This book remained the leading text on metallurgy and mining for many years. In Great Britain Roger Bacon, who has been described as the philosopher of industrial science, dreamed of using existing colleges of his time as scientific schools, a plan described in his diary in 1608. One of the oldest engineering schools was the Ecole des Ponts et Chaussées, established in France in 1747 for the purpose of training civil engineers. It was followed by others in that country, such as the Ecole Polytechnique in 1795. One of the earliest engineering schools in Britain was a college established by the East India Company near Croydon in 1809. In Germany technical high schools created early in the nineteenth century were transformed into engineering colleges, one of the earliest being that founded in Darmstadt in 1822.[10]

The main reason for the replacement of informal means of education with schools, colleges, and technical institutes was the growing complexity of engineering. For example, in the earliest times the choice of materials for a project such as a bridge was limited to a few common items readily available in the area, the principles upon which the bridge was built could be taught as strict rules for the neophyte to follow under the guidance of a more experienced engineer. In contrast, the selection of materials and the design problems of creating the relatively simple loco-motive engine in the nineteenth century was typical of a more complex profession, in which engineers could no longer safely rely on the repetition of what was done by their predecessors nor on the advice of an older mentor who may never have been involved in such a project. Engineers had to have training in the basic principles of physics, the properties of engineering materials, the use of mathematical methods of problem solving, and hosts of other subjects. The most efficient means of training engineers was the establishment of formal schools operated for that purpose.

The development of engineering education in the United States followed a pattern similar in many ways to the evolution of educational techniques and institutions in Europe. A perceptive article by Rae points out three sources of engineers in America in the first half of the nineteenth century, apart from the slow process of establishing academic schools and institutes. First was the practice of importing engineers from Europe, such as the German-born John Roebling, who designed the Brooklyn Bridge.[11]

The second source was the U.S. Army. As early as 1794 it began training a limited number of engineers, a responsibility taken over by the U.S. Military Academy at West Point in 1802. The Corps of Engineers was known as the elite branch of the army for a long time. Included in the ranks of the corps were such notables as Robert E. Lee and George B. McClellan. Many en-gineers went on to become well-known civilian engineers after leaving the service.

Rae listed as the third source self-education or a system of apprenticeship. As an example of the latter method, DeWitt Clinton, promoter of the Erie Canal, took three local landowners who had previously had some training in surveying and had them coached so they could assume supervisory duties in connection with the building of the canal.

Although early engineering education in America was dependent upon the previously described methods of training, since approximately the middle of the nineteenth century engineering education has been chiefly the responsibility of colleges, universities, and special institutes of higher education. The earliest civilian school established in this country was Rensselaer School, later to be renamed Rensselaer Polytechnic Institute, at Troy, New York; it offered civil engineering courses as early as 1831, and in 1835 created a department of engineering. However, not until 1886 could its students obtain degrees in mechanical engineering. Other early courses in engineering were given at Union College (1845), Yale University (1847), Harvard University (1848), Cooper Union for the Advancement of Science and Art (1859), and a few others. Near the end of the Civil War two other schools were added—Columbia University's School of Mines (1864) and Massachusetts Institute of Technology (1865). Thus it was gradually becoming easier for students interested in various types of engineering to find a college or institute teaching the desired program.[12]

However, a study of the educational situation made in 1866 shows that Rensselaer was the only school at that time providing a four-year course in engineering, with the others requiring one or more years of a general course before offering technical studies.[13] After the Civil War there was a noticeable increase in the number of colleges and universities granting degrees in engineering; a study published in 1885 listed twenty-two U.S. institutions offering such programs.[14] The period from 1870 to 1885 was described as one that "witnessed a remarkable growth in [engineering] enrollment and a differentiation of the profession into its main branches." The writer went on to say, "The increase of engineering graduates from less than 900 in the pioneer period before 1870 to over 3,800 in the decade of the eighties was due in large part to the publicly supported departments." He was referring, of course, to the increase in number and influence of state-supported colleges in many areas of the country.[15]

At this time there are several hundred public and private institutions in this country which offer undergraduate and graduate degrees in engineering, with a score or more of branches of engineering in which to specialize. However, one trend in recent years has been the development of curricula in which the old boundary lines between the branches of engineering are becoming less distinct. The current approach is more science oriented and increasingly multidisciplinary.[16] One example of this is the training evolved in recent years for environmental engineers, who must be knowledgeable in many subject areas— civil engineering, meteorology, biology, economics, and others. Prospective engineers in most programs must show proficiency in physics, chemistry, and mathematics, as well as in the subjects more closely associated with engineering per se. A study made by Emmerson shows that many schools now include courses dealing with the relationship of engineering to society, as well as a code of ethics for engineers to follow in their careers.[17]

TECHNICAL LITERATURE

Publications of interest to engineers existed in the earliest periods in which engineers were known to have worked, including clay tablets and manuscripts from ancient civilizations. Many of these have been preserved in scholarly libraries of the world, along with materials dealing with the pure sciences. Thornton and Tully have traced the origins and known collections of such early forms of technical information used by engineers before the invention of printing. By 1470, just a score of years after the advent of printed books, the first two scientific works had been published.[18]

However, it was to be nearly two hundred years before the initial publication of what is probably the most important type of literature for scientists and engineers: periodicals. The first scientific journal—the *Journal des Scavans*—was issued in France in 1665, to be followed a few months later by the *Philosophical Transactions* of the Royal Society of London, a title still being published. For quite a while issues of the *Transactions* included articles of interest to engineers, such as a paper on how to temper steel, a description of early iron works, or a discussion of a pump for use in removing water from mines. Journals of interest exclusively to engineers included one seventeenth century title, *Collection for improvement of husbandry and trade* (1691). After that, no other early titles appeared until 1823, when the British *Mechanics Magazine* began, with such features as accounts of new processes, practical applications of science, and reports on the state of the art of various technical fields. America made its entry into this field in 1825 with *American Mechanic's Magazine,* which subsequently became *Journal of the Franklin Institute* and which is still being published. A title devoted to the broad scope of all areas of technology, *Engineering,* started publication in 1866. By the end of the nineteenth century the number of engineering journals had begun to mushroom, although a number of them had a short life. In 1884 an index to engineering periodicals entitled *Descriptive Index of Current Engineering Literature* began to be issued; it was the forerunner of *Engineering Index,* a publication of great importance to the field of engineering.[19]

Thus by the turn of the century a significant number of serials and books devoted to engineering had developed. In addition to the traditional published literature of that era there was a growing body of technical information found in patents, a form of publication of particular interest to engineers. Several decades after the early 1900s there was a remarkable increase in the annual number of patents issued. Another type of engineering literature, technical reports, began to appear in significant numbers during World War I, but not until World War II did report literature gain prominence and begin an escalation in the quantity produced. The engineering profession is now dependent upon its literature for descriptions of new techniques and processes, listings of numerical data, formulas and principles, and other informational features. The printed literature has been augmented by a host of computer-readable technical databases, which provide new avenues of information retrieval. Computerized searching has proved to be a very efficient means of keeping abreast of technical information.

ENGINEERING SOCIETIES

Of great importance in the growth of any profession is the existence of professional organizations or societies which serve to gather the practitioners of a particular discipline and to spur them on in their profession by means of meetings, society publications, and special membership requirements. This holds true for engineering, although engineers had been at work at their projects for many centuries before engineering societies were ever dreamed of. Nevertheless, the growth of engineering as a profession was closely tied to the gradual emergence of societies for engineers.

One of the most important events in the development of science and engineering occurred in Great Britain in 1662, when Charles II granted a charter for the formation of the Royal Society of London. One of its founders was the distinguished scientist Robert Hooke. In his draft of the preamble to the society's statutes he wrote: "The business of the Royal Society is: To improve the knowledge of natural things, and all useful Arts, Manufactures, Mechanick practices, Engynes and Inventions by Experiment."[20] Engineering, therefore, was very much a part of the society's scope of interest in the early years of its history, although it is now known as a scientific society.

The first English engineering society was formed in 1771 when the Society of Civil Engineers was established under the leadership of John Smeaton (1724-92), said to be the first Englishman to call himself a civil engineers.[21] The group, sometimes known as the Smeatonian Society, became one of several professional associations established on local levels for engineers responsible for supervising road building, the draining of mines, and similar technical tasks. Briefly after Smeaton's death the society lapsed into inactivity, to be revived in 1793 by Sir John Rennie and W.C. Mylne, both engineers and builders of note.

The founding of the next engineering society which was to survive did not take place until 1818, when the Institution of Civil Engineers was formed. Within two years Thomas Telford (1757-1834), a famous canal engineer of the time, became president of the group.[22] He helped the institution acquire its own headquarters in the Strand area of London and later gave it his own personal library, including a collection of French engineering books. The institution soon considered publishing its own journal, but not until 1836 did it actually begin to issue its *Proceedings* and *Transactions*. The titles varied over the years, but publication has continued since that time.

Since civil engineering probably has had the longest span of existence of all types of engineering, in view of its practitioners' involvement in such basic projects as building roads and bridges, it is not surprising that this was the first branch of engineering to form societies in England. Historically, mechanical engineers were probably next in rising to prominence in any one country because of their work in designing machines, motors, and the like. The mechanical engineers formed a society in Great Britain relatively soon after the civil engineers. In 1847 the Institution of Mechanical Engineers was formally organized, with the noted engineer George Stephenson, designer of locomotives, as its first president. This society was said to have been founded largely as a result of the refusal of the civil engineers to award Stephenson membership in their institution unless he submitted evidence of his skill as an engineer, which to his colleagues seemed an indignity for a person of his status.[23] After that date there was a steady stream of engineering societies formed in Great Britain. At least fourteen were established before 1900, including groups of gas engineers, mining engineers, marine engineers, heating and ventilating engineers, and water engineers.[24]

Activity among engineers in the United States lagged behind that of the more mature culture found in Great Britain at that time, but even so there were some surprisingly early attempts to form societies in the more rough-hewn background of eighteenth century American life. For example, Wisely, in his thorough study of the civil engineering profession, cited the Society for Promoting the Improvements of Roads and Inland Navigation as having probably been formed in 1789. It was still in existence as late as 1826. Wisely reported unsuccessful attempts to form national civil engineering organizations in 1836, 1838, and 1841. In 1848 the Boston Society of Civil Engineers was founded; it survived until 1947, when it merged with the national organization.[25]

In view of the difficulties encountered by the would-be founders of a national organization for civil engineers, it is a tribute to the perseverance of a few leaders that they kept trying. They finally achieved success when a few men gathered on Friday evening, 5 November 1852, at the office of the Croton Aqueduct Department in New York City to form a new organization. Alfred

W. Craven, chief engineer of the Croton Aqueduct, presided at the
meeting, which was followed the next month by the adoption of
a constitution and by the first meeting for members on 5 January
1853. The group was organized as the American Society of Civil
Engineers and Architects, the name it used until 1868, when it
dropped the last two words. Its constitution stated the purpose
of the society to be "the professional improvement of its members,
the encouragement of social intercourse among men of practical
science, the advancement of engineering in its several branches,
and of architecture, and the establishment of a central point of
reference and union for its members."[26]

A circular issued at that time made it clear that the society
was interested in having applicants from those in other branches
of engineering: "Civil, geological, mining, and mechanical
engineers, architects, and other persons, who, by profession, are
interested in the advancement of science, shall be eligible as
members." It went on to say that the union of the different types
of engineering specialties and architecture "will be attended by
the happiest results," and predicted that the membership of
diverse types of backgrounds would tend "to quiet the unworthy
jealousies which have tended to diminish the usefulness of distinct
societies formed heretofore by the several professions for their
individual benefit." This was to be the attitude held by ASCE
officials for many decades—that their organization welcomed
engineers from all branches of engineering. Library facilities
were also mentioned in the circular, as follows: "Among the
means to be employed for attaining these ends shall be periodical
meetings for the reading of professional papers, and the discussion
of scientific subjects, the foundation of a library, the collection
of maps, drawings and models, and the publication of such parts
of the proceedings as may be deemed expedient."[27]

Meetings of the group continued up to 2 March 1855, after
which there is no record of further activity until 1867, when
meetings were finally resumed and the minutes of the 1855 meeting
were at last approved. An explanation of the long lapse was given
in 1867 by ASCE President James P. Kirkwood, who said it was
caused by the fact that "no place was established to which members 'could point as evidence of our being something more than a

parchment Society' and that in order to maintain a bond between all classes of members it was necessary that papers should be printed and disseminated to those who could not be present at the meetings."[28] Other reasons for the interruption of activity include the Civil War's disruption of life in general and the fact that many members and officers of the society served in the army during the war.

Next in the evolution of engineering societies of particular interest to this study is the development of the American Institute of Mining Engineers, which began on 16 May 1871, when three engineers distributed a circular and placed notices in publications calling for a meeting in Wilkes-Barre, Pennsylvania. Unlike the civil engineers, their objectives, adopted at the 16 May meeting, were explicit about the type of engineers they were seeking as members. They stated: The objects of the American Institute of Mining Engineers are to enable its members, comprising Mining Engineers and other persons interested in mining and metallurgy, to meet together at fixed periods for the purpose of reading papers upon and discussing subjects which have for their aim the economical production of the useful minerals and metals, and the safety and welfare of those employed in these industries, and to circulate among its members, by means of publications, the information thus obtained.[29] The institute headquarters was located for many years in buildings at Lafayette College in Easton, Pennsylvania, and not until approximately 1900 did it move to New York City.

Less than a decade after the mining engineers formed their society, the mechanical engineers took a similar step. Early in 1880 an invitational letter was sent to about fifty engineers, notifying them of a meeting to be held in New York City in February. Some forty persons attended, "some coming from as far west as Ohio," and letters of approval of the idea were sent by over a dozen who could not attend. In view of the response the attendees decided to form the American Society of Mechanical Engineers.[30] The formal organizational meeting took place at Stevens Institute of Technology in Hoboken, New Jersey, on 7 April 1880. The eighty people present included blast-furnace engineers, railroad designers, steam engineers, engineers for manufacturers of sewing machines, metallurgists, and the like.[31] The meeting adopted the following statement: "The objects of the AMERICAN SOCIETY OF MECHANICAL ENGINEERS are to promote the Arts and Sciences connected with Engineering and Mechanical Construction, by means of meetings for social intercourse and the reading and discussion of professional papers, and to circulate, by means of publications among its members, the information thus obtained.[32] Sinclair's thoughtful history of ASME's first century emphasized the importance of the role played by certain dedicated engineers in its creation and growth.[33]

The American Institute of Electrical Engineers, the last of what were to be the first four Founder Societies of the Engineering Societies Library, was established in 1884. The first sign of activity was the traditional one of sending out a circular urging the formation of a society for electrical engineers, to be modeled upon the previously formed American engineering societies, and to include a library.[34] An organizational meeting was held in New York on 15 April 1884, with the first regular meeting held on 13 May. Having no quarters of their own, the electrical engineers met in the building of the American Society of Civil Engineers. Two well-known Americans were elected vice-presidents at the May meeting, Alexander Graham Bell and Thomas A. Edison.

At that meeting the group adopted its "Rules," which included the following statement of purpose:

> The objects of the AMERICAN INSTITUTE OF ELECTRICAL ENGINEERS are to promote the Arts and Sciences connected with the production and utilization of electricity, and the welfare of those employed in these industries; by means of meeting for social intercourse, the reading and discussion of professional papers, and the circulation, by means of publications among its members and associates, of the information thus obtained.[35]

3

Development of Technical Libraries

The development of engineering societies was a natural outgrowth of the increasing need of engineers to confer with their peers and to keep abreast of new developments, as engineering gradually became more complex in its nature and more dependent upon an ever-increasing body of professional literature. The latter development, the emergence over the years of a vast array of technically oriented publications, in turn led to the establishment of technical libraries, which became vital organizations in the process of making scientific and engineering information readily accessible to scientists and engineers. A number of technical libraries—including public, academic, industrial, and society-sponsored libraries—developed in New York City late in the nineteenth century.

This chapter will present a brief account of the founding and development of the libraries of the four societies which later became the original sponsors of the Engineering Societies Library, as well as a review of the establishment of technical libraries in New York City and other regions of the United States, with greatest emphasis on the era near the turn of the century.

FORMATION OF SOCIETY LIBRARIES

By 1884 the four engineering societies later to become the original Founder Societies of the Engineering Societies Library were all in existence, established at different times, with separate goals and separate locations. Each society developed its own library, each different in purpose and size. An essay by William Tornow provides a detailed account of the establishment of these libraries[1], but a brief description of the process and of the resulting libraries is worth presenting here as a background for later events.

American Society of Civil Engineers

The fact that the American Society of Civil Engineers (ASCE) was the first of the four societies to be formed makes it understandable that it was the first of the four groups to have its own library. Soon after the postwar resumption of meetings in 1867 the society adopted its constitution, which specified the duties of the librarian and the library committee, as follows:

> Article XI. It shall be the duty of the Librarian to take charge of the Library of the Society, and to see that all books are marked with the name of the Society, numbered, and recorded in a catalogue. In respect to the management of the Library, he shall conform to such regulations as may be prescribed by the Society or by the Board of Directors.
> Article XIV. The Library Committee shall have supervision of the rooms and Library of the Society, and shall apply to the purchase of books, or other articles of permanent value to the Society, such sums as may be appropriated.[2]

These rules seem to have placed the librarian in the role of a clerk, with the intellectual assignments given to the library committee. However, one should keep in mind the relatively low status which librarianship had at that time, with few standards and virtually no formal training available for would-be librarians. These rules were an effort to insure that certain basic processes would be carried out.

Little or nothing was done towards actually establishing the library until 1873, when a special committee began to prepare a report which set forth the principles by which the library should be operated. Its report, finished in 1875, emphasized the type of collection to be acquired:

> The library...should contain the literature of rational and applied science, constructive art and technology; all that has been, or may from time to time be published, relating to the history and prosecution of engineering; the maps and profiles of every canal and railroad, their complete reports and those of municipal and state departments; descriptions of private and miscellaneous works; statistics of the material resources and development, the wealth, manufactures and commerce of countries; standard works of reference in science and art, and lack nothing published anywhere, in our own or other tongue, that in a library may aid the student or accomplished engineer seeking professional knowledge.[3]

This report introduced an ambitious goal, one that called for an exhaustive collection of books, journals, reports, and statistics on everything ever published in any language on civil engineering. While one can admire the desire of the founding fathers of the society to have an excellent library, it is apparent that they had no concept of the size of staff and the amount of space it would require to develop a library of this caliber and size.

Actually the library had a small staff and a relatively small area in which to operate. For example, the head of the library bore the title of "Secretary of the Society and Librarian." Thus the library had only a part-time head from the beginning, and kept this arrangement throughout its existence. It occupied a space which, at the most, had room for 50,000 volumes.[4]

A collection of the sort described in the committee's report would have been more appropriate for the Library of Congress, and even that library would probably have reduced in some way the great quantity of literature called for in the statement of collection goals. For example, it might have decided to collect selectively in the area of canal and railroad company reports rather than to attempt to obtain all of them. Likewise, in regard to the languages in which items were published, it might have decided to concentrate on certain major languages and to collect very selectively in others.

The actual growth figures for the collection show that acquisitions were at a much more modest level than the initial grandiose plan. By 1873, after one year of operation, the collection consisted of approximately 3,400 items, most of which were acquired as gifts. Soon thereafter the library had a card catalog and rules for the circulation of materials. By 1892 the collection contained 16,000 items, which had increased to 22,000 by the time the society moved to its new building on West Fifty-seventh Street in 1897.[5]

One of the library's employees recalled that at the turn of the century the collection included several fields, such as sanitary engineering, mechanical engineering, and electrical engineering, in addition to civil engineering. Also included were annual reports, such as those of railroads and municipal waterworks. However, it was far from the all-encompassing collection for which the committee hoped. In regard to its services, it was a reference library, open not only to ASCE members but also to the general public.[6] A most interesting point to note is that the library had been authorized by the ASCE board as early as 1880 to make searches on a fee basis for engineers requesting special assistance.[7] It is known that about a thousand searches were completed in the period 1902-17.[8] This is thus another piece of evidence that fee-based reference service on scientific and technical data has a much longer history than most librarians realize.

After the ASCE moved to its new home, its secretary-librarian, Charles Warren Hunt, felt that the collection was not well organized. In 1897 he developed a classification scheme of his own, which was used in the creation of a classified card catalog; it took from 1897 until 1899 to reclassify the collection.[9] The next step was the publication in 1900 of a 700-page book catalog of the collection; it listed approximately 16,000 titles, which represented total holdings of about 32,000 volumes, pamphlets, maps, photographs, and specifications. In 1902 a supplement to the book catalog of some 290 pages was published. Evidence of Hunt's interest in the library classification scheme used for the book catalog and the card catalog is seen in the fact that he devoted almost fifty pages to a reproduction of the classification system (as revised by two members of the library staff) in an article he wrote on the history of the ASCE.[10]

American Institute of Mining Engineers

The next society library formed was that of the American Institute of Mining Engineers (AIME). In this case, the catalyst which acted to bring about the creation of the library was the Centennial Exposition in Philadelphia in 1876. The AIME decided to maintain a set of rooms at the exposition in which could be found "technical periodicals and books of reference."[11] At the close of the exposition society members generally agreed that it would be unfortunate to scatter the collection, so the committee in charge of the rooms tooks steps to keep the material together and to solicit gifts to add to the nucleus. By 1877 they had assembled what was considered at the time at least to be "a very valuable and complete library of reference."[12]

For the next two years the AIME library was housed in the office of a member of the society, Dr. Thomas M. Drown, on the campus of Lafayette College in Easton, Pennsylvania, where he was a faculty member. Unfortunately a disastrous fire swept the building containing his office and only the official records of the institute were saved, the library being totally destroyed.[13] Little is known about the rebuilding of the library after the fire, but when the institute moved to New York City around 1900 the library was mentioned as being part of the new headquarters.[14]

The library grew steadily, enriched by numerous gifts or judicious purchases of collections or whole private libraries. The institute's bylaws included the following section regarding the library:

> X. Library Committee.
> The Library Committee shall be the custodian of all books in the Institute Library and of additions thereto: also of all back numbers of the *Transactions* of the Institute. It shall, on the first day of May, of each year, receive from the Secretary of the Council, and receipt for same to him, all the volumes of *Transactions* for the

preceding year, not then distributed by said Secretary. It shall cause to be kept, under the direction of the Assistant Secretary, a catalogue of all books in the Library and an account in ledger form of all volumes of *Transactions* in its custody, in which shall be charged to it all volumes delivered to it, and in which shall be credited all volumes taken from its custody for sale or for any other purpose. The receipts from the sale of any volume of *Transactions* taken from the custody of the Library Committee shall be credited to the Library Committee on the books of the Treasurer, and be devoted to the general purpose of the Institute.[15]

This set of library guidelines was even more clerical in tone than those of the ASCE library and gave the responsibility for daily operations to the library committee rather than to the librarian. However, these conditions did not deter the library's first known librarian, Miss L. Elizabeth Howard, who served in this position from 1901 through 1908. During this period she established a card catalog for the collection, obtained Library of Congress cards for appropriate materials, classified volumes according to the Dewey Decimal System, used a distinctive color coding system in binding the periodicals to indicate their subject scope, and systematically sought out missing volumes for their runs of serials.[16] While this by no means exceeds what is commonly done in libraries today, it was many steps beyond the record-keeping concept set forth in the formal guidelines.

American Institute of Electrical Engineers

The next library founded was that of the American Institute of Electrical Engineers (AIEE). One of the reasons for its establishment was the difficulties the secretary of the institute reported on in 1885; he found that he was hampered in setting up exchanges of publications with Canadian and European scientific and technical societies by there being no institute library.[17] The institute's own library began soon after that, although in an informal way—mostly in the form of gifts of literature which were turned over to the institute's secretary for safekeeping. By 1900 progress had been made to the point where a library committee for the institute had been appointed, and $500 had been appropriated for the purchase of two book stacks and for binding some of the periodicals which the institute had received on exchange. It is also known that the collection at that point had grown to several hundred volumes.[18]

American Society of Mechanical Engineers

The last of the four libraries to be established was that of the American Society of Mechanical Engineers (ASME). Although the society was formed in 1880, no attempt was made during the next three or four years to start a library. In 1883 requests were made for donations of literature, and the results were so good that a committee was formed that year to investigate the desirability of establishing a library for the society.[19] The committee presented a thoughtful report at the annual meeting of the society in November 1884, which contained seven recommendations regarding the foundation of a library. The report called for the establishment of a standing committee for the library (which was done in 1885); it urged that all members be invited to contribute funds or literature to the library; it recommended that permanent headquarters be obtained for the library; and it stated the opinion that all materials in the library should be available for loan if a file copy were retained.[20]

Within the next five years several recommendations in this report were implemented, with progress evident as early as 1885, when it was reported by the committee that the library had been formed. The list of accessions for 1885 was published, a practice which was to be continued for many years.[21] During this period the ASME headquarters was moved several times, finally ending up in May 1890 in the building formerly owned by the New York Academy of Medicine on West Thirty-first Street.

OTHER NEW YORK CITY LIBRARIES

When one considers other libraries in existence in New York City in the 1890s which had significant technical collections, it becomes evident that several different types—public, academic, society, and private libraries—were involved. Libraries for the public in the city were slow in their development relative to the situation in many other large cities in the United States. Ample comparisons between New York and other cities may be found in Harry Lydenberg's history of the New York Public Library; he describes the superior library facilties in such cities as Boston, Cincinnati, and Chicago.[22] Phyllis Dain, in her study of the founding and early years of the New York Public Library, says: "There was no extensive free public library system or large central circulating collection for its [New York City's] millions of inhabitants and workers, so many of them bent on advancement through self-education, that could compare with those of Chicago, Boston, or even smaller American cities, none 'worthy to be called a comprehensive free library.'"[23]

Both Dain and Lydenberg discuss the problems that limited the usefulness of the Astor Library and the Lenox Library to the public. The Astor Library, which had opened in 1854, was designed to be a free reference library, yet it had unusually short hours of service, closing at dusk for lack of either gas or electric lights, a practice that limited many researchers. The Lenox Library, opened in 1877, was restricted in the availability of its collections to the public. Users were required to obtain the permission of the superintendent of the library to consult materials, and even such limited use was not permitted until 1882. Some critics called the Lenox Library more a museum than a library, and both libraries were the object of satire in the contemporary press because of their lack of accessibility for the average user.[24]

The situation in New York City seemed certain to change by the 1890s when the disposition of Samuel J. Tilden's will, which had provided for the funding of a free public library, was finally settled. The Tilden Trust, created to carry out the will, was left with an endowment of about two and a half million dollars. There was a steady stream of applications from libraries, academic institutions, and other groups wanting to share the money, and there were differences of opinion among the Tilden trustees about the best use of the funds. Among the applicants was a New York City group known as the Scientific Alliance. As Dain described it, the alliance was "a confederation of the chief scientific societies of New York City that hoped to promote scientific research and popular scientific education in a way that the New York Academy of Science, a member, had not done." The alliance proposed that the Tilden funds be used to construct a headquarters building for the alliance, which would include a cooperative science library.[25]

A plea for this project may be found in a compilation of addresses delivered at the first meeting of the Scientific Alliance in 1892. The talk, by H. Carrington Bolton, proposed that a library devoted to pure and applied science be built: "It would become the headquarters of those engaged in pure research as well as of inventors and others seeking data as to the applications of science."[26] Bolton appended to his paper a directory of sixty libraries in the New York City area, out of sixty-eight queried, with data on their collection strength in the sciences and technology. Collectively they owned about two million volumes in all disciplines, but the percentage of scientific volumes in each library ranged from 5 to 100 percent. The ASCE library had one of the highest percentages of scientific materials in its collections; approximately 88 percent of its 15,000 volumes were technical in nature. Evidently the sixty libraries which took part in the survey accounted for about half the volumes in all libraries in the area, in view of a survey which showed that in 1902 the total collections of libraries in and around New York City amounted to just over four and a half million volumes.[27]

Despite the efforts of the Scientific Alliance, the Tilden trustees decided that the money would be used to create a public library covering all disciplines, not just science, and no role was allowed for the alliance as a participant in the merger of the trust with the Astor and Lenox Libraries in 1895 to form the New York Public Library. It was not until 1896, however, that the alliance finally gave up hope of a share in the Tilden Trust plans; it then decided to have an architect draw up plans for its own building. The building committee report showed that the alliance was hoping to have a five-story building large enough to house offices for member societies, and to include a thousand-seat auditorium and a library with a capacity for 200,000 volumes.[28] A list of the eight member societies of the alliance included no engineering groups, so that presumably the proposed library would not have contained engineering literature. The alliance never raised the funds for this ambitious building project, and the organization ceased to exist in 1907. It is interesting to note that within a few years the engineering societies in New York would be seeking ways to construct a similar type of building.

When the New York Public Library opened its new building at Fifth Avenue and Forty-second Street in 1911, it included a Technology and Patent Division, comprised of materials gathered from the Lenox and Astor libraries, as well as substantial collections acquired since 1896. These selections were made personally by Dr. John Shaw Billings, director of the library, with the assistance of the head of the division. Billings was an internationally known medical leader who had built the Surgeon General's Library in Washington into a great medical library. He and the chief of the technology division also selected technical books for the central circulation unit and the library's circulating branches, which numbered forty at that time. When the new building opened there were approximately 5,000 monographs on hand in the division. A contemporary article by William G. Gamble, first chief of the division, described its quarters as consisting of "five magnificent rooms." The division, like the other service units in the building, was operated as a reference library for research purposes, open to the general public. As Gamble put it, "The division is intended for the man who would 'get at the bottom of things'— the engineer, the inventor, the manufacturer."[29]

In addition to the Technology and Patent Division there was also a Science Division, which had a collection of about 4,800 monographs when the library opened. Billings gave this division the same assistance in collection building as he did for the technical unit.[30] The two sections remained separate until 1919, when they were merged into one department.[31]

A few years after the Astor-Lenox-Tilden consolidation, the figure of Andrew Carnegie became prominent in the library scene in New York. Carnegie had moved to that city in 1897, where he served on the board of the New York Free Circulating Library before it merged with the New York Public Library in 1901. In that same year he made a spectacular contribution in the form of

an offer of over five million dollars to build sixty-five branch
library buildings in all five boroughs of New York City, on condition that the city provide the sites and operating expense, the
latter to be at least 10 percent of his contribution. Needless to
say, his offer was accepted by the city and its three main public
library systems, New York Public Library, Queensborough Public Library, and Brooklyn Public Library.[32]

   In subsequent years various independent free circulating
libraries in Manhattan, Bronx, and Staten Island, which had been
receiving city funds, were given to understand that such aid would
eventually be cut off unless they merged with the New York
Public Library. Many did give up their independence and agreed
to consolidation, but some libraries decided not to join the public
system and thereafter operated without city aid. One such library
was that maintained by the General Society of Mechanics and
Tradesmen, which had been established in 1820. After losing
city funds it reduced its efforts to be a general public circulating
library. Dain described the situation as follows:

> It then concentrated on serving the purposes of the
> General Society, which had evolved into primarily an
> educational agency for technical training. The library
> could exist without municipal subsidies, in part because
> it was benefiting directly from Andrew Carnegie's
> benevolence. Carnegie had been a member of the Society
> for years and between 1899 and 1908 gave it $527,000,
> about two-thirds of which was used for building altera-
> tions that included the library and the rest for endow-
> ment.[33]

In 1901 the library had more than 105,000 volumes, which included
fiction and social science as well as technical subjects.[34]

   An additional library which provided the public with scientific
and technical materials belonged to the Cooper Union for the
Advancement of Science and Art, established in 1859 "with the
express purpose of improving the working classes, from whose
ranks he [Peter Cooper] had forced his way to wealth and influence...." The fifth annual report of the institution noted that the
library reading room was free to all to use "without tickets."[35]
By 1909 the fiftieth annual report showed that the Cooper Union

library had some 48,000 volumes, of which approximately 25 percent were devoted to science, applied science, and patents. The periodicals added that year included such titles as *Gleanings in Bee Culture, Barge Canal Bulletin,* and *Philippine Journal of Science* for technical readers, along with *Numismatics, Violinist,* and *Gaelic American* for the nonscience readers.[36]

In another sector of New York's educational world, Columbia University had an early start in collecting scientific materials. It offered courses in the sciences at the time of its opening in 1754 and must have had literature on such subjects when its library opened in 1761. Its School of Mines opened in 1864, which was an impetus to the building of a collection in that subject. By 1901 Columbia had separate libraries for chemistry and physics; libraries for geology, psychology, and zoology-botany opened in 1912.[37] By 1915 the Columbia libraries held approximately 30,000 volumes on technical subjects.[38] A cooperative project was carried out by Columbia and the New York Public Library in the late 1890s: preparation of a union list of serials which included various science and technology sections, such as one for chemistry.[39]

Still another college-level library in that era was at the Pratt Institute in Brooklyn, which opened in 1888. In 1905 the annual report contained a statement about the opening of the Applied Science Reference Room, which included books, periodicals, patents, and trade catalogs.[40]

The contribution made to the scientific library world by special business and technical libraries is assessed in Anthony Kruzas's historical survey of science-technology libraries in the New York region in the period around 1900. He lists the engineering firm of Ford, Bacon and Davis (1894), American Telephone and Telegraph Company (1907), the German Kali Works (1890), Consolidated Gas Company (1906), and the General Electric Research Laboratories at Schenectady (1901).[41]

These pioneer special libraries may well have acted as a spur for the creation of similar libraries in other commerical organizations, since Kruzas shows that of 114 company libraries formed before 1910, 44 percent of them were organized in the period 1905-09.[42] Yet, despite the growth of special libraries in that era, when the Special Libraries Association was formed in 1909 partly as an alternative to the American Library Association, of the twenty-six people present at the first organizational meeting all but nine were public librarians.[43] In considering the role played by these special libraries in New York, it should be kept in mind that most of them were restricted to serving the employees of their sponsoring organizations and hence were normally of little use to the general public.

One more type of technical library existed in the New York area—a library sponsored by a society of scientists and engineers. The Chemists' Club library, sponsored by a society made up of chemists and chemical engineers working in the metropolitan New York area, was such a library. It was established in 1898 and was open to the public. By 1915 the library was subscribing to nearly 450 periodicals, with about 18,000 volumes. It concentrated on chemical engineering as well as chemistry.[44]

Thus at the turn of the century New York City possessed several libraries which had fairly significant collections of scientific and/or engineering literature. Many were open to the public, but few allowed books to circulate. There was virtually no effort made to avoid duplication of materials through cooperative agreements; each library went its own way in building its collection.

TECHNICAL LIBRARIES IN OTHER AREAS

Important technical libraries were developing outside the New York area at the turn of the century, some being open to the public. One public library of particular importance was founded in Chicago at about the same time as the New York Public Library was created. It was established according to the terms of a will of a prominent wealthy Chicago philanthropist and manufacturer of railroad supplies, John Crerar. In response to a legal challenge from Crerar's relatives, the Supreme Court of Illinois declared the will valid, which made some two and a half million dollars available in 1894 for the library. In keeping with Crerar's strong moral and religious principles, the will stated:

> I desire that the books and periodicals be selected with
> a view to create and sustain a healthy moral and Chris-
> tian sentiment in the community and all nastiness and

immorality be excluded. I do not mean by this that there shall be nothing but hymn books and sermons, but I mean that dirty French novels and all skeptical trash and works of questionable moral tone shall never be found in this library. I want its atmosphere that of Christian refinement, and its aim and object the building up of character, and I rest content that the friends I have named will carry out my wishes in those particulars.[45]

The directors named in Crerar's will executed their charge conscientiously. In 1894 a temporary committee of directors made a serious study of existing Chicago libraries to determine the fields of interest best suited for the Crerar library. More than a score of libraries responded to their survey, which showed, for example, that the Chicago Public Library, established in 1872, emphasized the humanities; while the Newberry Library, an endowed free reference library which opened in 1893, concentrated on music, medicine, history, and bibliography. The Crerar directors also asked a number of high-ranking librarians throughout the United States for suggestions about the type of library to establish. Several respondents wrote in favor of an applied science and technology library, while others suggested topics ranging from agriculture to railroads. At a meeting held in January 1895, the directors issued a tentative statement on the scope of the new library:

A free reference library and a library of science, containing books which may be classed under the head of popular science, practical science and applied science... and that there be departments of social science, architecture, civil and mechanical engineering and astronomy. The policy of the library is to limit the number of departments to such as can be fully supplied and maintained by the resources of the corporation.[46]

Later in 1895 the final decision was made to allocate to the Crerar Library the subject fields of "Philosophy, the Physical and Natural Sciences; the Useful Arts (Technology); the Fine Arts in part; Sociology and Economics." Obviously there was little danger of racy books being selected with such collection goals. Medicine was to remain at the Newberry Library, and the Chicago Public Library was to concentrate on books for the public which were "wholesomely entertaining and generally instructive...."[47] These allocations were to undergo numerous changes in succeeding years, notably when Crerar took over coverage of medicine from Newberry and gave up the social sciences, but the collaboration of the three libraries in establishing goals for their collections, an unusual example of urban library cooperation, continues to this day.

Crerar's first librarian was Clement W. Andrews, a chemist who had been in charge of the chemistry library at the Massachusetts Institute of Technology and who later headed the entire MIT library system. Under his leadership the Crerar Library made rapid growth in its early years. In 1896, a few months before its spring opening, it had about 11,000 volumes accessioned and it subscribed to approximately 170 journals. Two years later the collection had grown to 29,000 volumes in science and technology, 13,000 in the social sciences, and about 12,000 general works. By the time Andrews retired in 1930 the collection had grown from nothing to more than half a million volumes.[48] Andrews had made several major decisions in his term: one was to establish a classed catalog based on the Dewey Decimal Classification; another was to plan a separate library and then move it in 1920 to a new location across the street from the Chicago Public Library in the heart of the city.[49]

Still another public library with strong materials was being formed in the 1890s, this one in Pittsburgh, the area in which Andrew Carnegie had lived for many years. As a youth in the 1850s Carnegie had worked there as a telegraph operator, then in later years made his fortune in the steel industry. During his boyhood he found his love of reading encouraged by the availability of a local private library.[50] To show his appreciation of the value of libraries, Carnegie subsequently gave millions of dollars to establish public libraries all over the United States. In 1881 he offered a quarter of a million dollars to Pittsburgh for the formation of a free public library; in 1890 he increased the offer to one million dollars in addition to an annual gift of at least forty thousand dollars. The increased size of the gift was partially due

to his desire to see established not only a library, but also a museum
and related cultural units. The offer was finally accepted by the
city, and work began in 1890. The Carnegie Library of Pittsburgh
was opened to the public in 1895 with a collection of about 16,000
volumes.[51]

It should be noted that Carnegie's aid to Pittsburgh was one
of his earliest acts of philanthropy as far as libraries were concerned. From 1886 to 1896 he gave nearly two million dollars for
public buildings; he later spent money at a much faster rate in
what one biographer called Carnegie's "wholesale" period in
contrast to his "retail" period before then. From 1896 to 1919 he
gave over thirty-nine million dollars to libraries.[52]

The annual report of the librarian of the Carnegie Library
at the end of the first year, published in 1897, included a plea for
funds to establish a strong technical collection of books and
periodicals. One reason for this was the obvious importance of
industry, notably the steel industry, to the city and the subsequent
value of such a collection to users seeking technical materials.
The report clearly showed that the library was being used exten-
sively; in 1897 the library served 270,000 people, a number equal
to Pittsburgh's entire population at that time.[53] Carnegie promptly
offered the library $10,000 for increasing the technical collections,
and the library later became nationally known for its strong holdings in science and technology, especially to the iron and steel
industries. In 1902 the Carnegie Library of Pittsburgh became
the first public library in the United States to have a separate
department for its science/technology service, one that enjoyed
heavy use from the beginning.[54]

In addition to the development of public libraries offering
strong collections and adequate service in science and engineering subjects, a number of company-sponsored libraries specializing
in these disciplines sprang up all over the United States around
1900. Kruzas cited such examples as pharmaceutical libraries in
Chicago (1888) and Indianapolis (1881), as well as two technically
oriented libraries in Boston, at the Arthur D. Little Company
(1886) and the engineering firm of Stone & Webster (1900).[55]

In summary, it is apparent that the increase in science/
technology libraries in the United States at the end of the nine-

teenth century and continuing into the twentieth was not confined to any one region. However, the greatest number of such libraries was located in urban areas, close to the industries, research organizations, and business centers which needed ready access to technical information. New York City, with its concentration of business and technical enterprises, had a number of libraries with strong technical collections. Another reason for such libraries was the presence of men like John Shaw Billings, whose scientific background gave him a personal insight into the need for outstanding scientific and technical collections in an institution such as the newly created New York Public Library, which aimed at excellence in all disciplines.

One additional factor was the high concentration of engineers in and around New York City, as evidenced in part by the number of engineering societies which established their headquarters in the city during the last part of the nineteenth century. A study of the geographical distribution of civil engineers in the country in 1880 shows that 13 percent of them lived in New York State, the highest percentage for any state.[56]

In New York City the advantages to independent libraries of joining a consolidation of libraries was brought home in the public library sector in the early 1900s, after Carnegie's grants were used by public library officials as a means of bringing individual libraries into the New York Public Library system. The array of formerly independent free circulating libraries was considerably diminished as one after another joined the consolidated movement.

An additional stimulus to the creation of new libraries in New York City was the presence of Andrew Carnegie, for many years an inhabitant there. His spectacular gift to the New York Public Library in 1901 left no doubt in anyone's mind of his willingness to make impressive gifts to support libraries. A library that would concentrate on technical subjects would have a special appeal for the former steel manufacturer, who owed his fortune to the management of technical equipment and complicated processes, and who had supported such a library already in Pittsburgh.

4

Creation of the Engineering Societies Library

The turn of the century has been seen as a period in which a combination of conditions and attitudes favorable to the creation of a technical library serving several major engineering societies had developed. This chapter describes the events spanning more than fifteen years, which led eventually to the establishment of the Engineering Societies Library.

The first recorded effort in this direction took place in 1885 when several engineers stated the need for the creation of a joint library for their societies. One of the pioneers in this movement was Henry R. Towne, a member of the American Society of Civil Engineers. At the January ASCE meeting that year he made the following statement:

> I have participated in the discussion concerning this matter with the members of two other societies—the Mechanical Engineers and the Mining Engineers—and...it has been with the hope that at some time in the future the three engineering societies now existing, with possibly at some time, the addition of a fourth, the Electrical, may unite in having a central headquarters, presumably in this city [New York], where a library shall be located, common to all of the societies, and where also they may unit in a building in which all can have suitable places for meeting....

His plea supported a motion that had been made to appoint an

ASCE committee of five members "to confer with committees appointed by kindred societies, for the purpose of devising and considering a plan for creating a library for the joint use of the organizations represented by the committees in conference; said plan to be reported to the society for such action as may be desirable.[1] An attempt by Towne to broaden the motion to include studying the concept of a headquarters building for the societies to share was defeated, but the original motion (to study the idea of a joint library) then passed.[2]

It is clear that there were several reasons why these engineers favored a joint library and headquarters building. One was simply the desire to create an adequate library capable of supporting members' reference work; another was to avoid a waste of funds due to duplicate purchases of literature by different society libraries, when one set would have sufficed; and finally, they wanted to house the library in a joint headquarters building where engineers could meet their friends and make new ones among members of other societies. It is rather surprising that so many years were to pass before this goal was achieved, in view of the clarity of the logic propounded by its advocates. The record shows that not every member at the meeting spoke in favor of a joint library, but there was a general mood of willingness to have it investigated. It is interesting to note that the ASCE was apparently the first of the four societies to propose the study, yet ironically it was to be the only one of the four that did not unite in a common building and library plan when the other three did.

The idea of a joint library was supported by the other societies in 1885. For example, at the annual meeting of the American Institute of Electrical Engineers the acting secretary, T. C. Martin, spoke in favor of the plan, which, in his opinion, was to consider not only a joint library but also a joint headquarters building. He noted that the AIEE Council had appointed members to take part in the work of the joint committee formed by the four societies.[3] Later in the meeting the chairman of the assembly, R. R. Hazard, reported that some of the discussions had centered on a plan "to get one of the large estates like the Astor or Rheinlander to put up a large fire-proof building to accomodate [sic] twenty-five or more societies which have common purpose, scientific or artistic, for preserving their records, their libraries and their specimens, to make a common rallying point for the scientific and artistic world in this city, which is the commercial centre of the United States."[4]

In the eyes of the AIEE officers the project was even broader than considering a joint home for engineering societies; the chairman mentioned the inclusion of artistic societies as well as "the Historical and Geographical societies." One reason why a joint headquarters may have had such an appeal was the fact that the AIEE at that time had no permanent home of its own, with all meetings held in the ASCE building, thanks to the courtesy of John Bogert, secretary (and librarian) of the ASCE.

While the joint committee was in existence there were editorial comments on the idea of a joint library in engineering journals. For example, a plea for the project appeared in 1887 in *Electrical World*: "Why should not the Civil, Mining, Mechanical and Electrical Engineers and the societies of the Architects and the Chemists unite in making a home of the arts and sciences sufficient to their needs and creditable to the city? With their combined resources they could soon have a well-arranged and a well-catalogued library...."[5] However, despite the joint committee's efforts, no progress was made in the next three years towards the preparation of a viable plan for a joint library, nor for a joint headquarters building. In 1888 the ASCE delegation to the committee was discharged, as were the representatives from other societies.[6] The need for a joint library was evidently not great enough at that time to overcome the obstacles in the way.

Within seven years, however, another effort was made, this time by an engineer named William D. Weaver, who wrote to Carnegie in 1895, proposing that the societies have "a combined headquarters and common library...." Weaver went on to point out the need for "a friend of the societies" who could provide funds for such a project.[7] Carnegie's answer was favorable, but it was not until December 1896 that Weaver replied to him. Carnegie's next response seemed rather cool. His secretary, P. Finnegan, wrote: "the matter about which you write is one which the members of the respective bodies themselves must settle."[8] Unfortunately, Weaver misinterpreted this letter as a sign of lessened interest on the part of Carnegie. Weaver wrote later that he had not realized that while Carnegie felt that engineers should participate in fund raising, he approved of the overall idea.[9]

An event which later played a significant role in the creation of a joint library involved the desire of an American engineer to acquire for the United States a major British collection of science literature. The collection had been assembled by Josiah Latimer Clark, a prominent British electrical engineer and inventor in the field of telegraphy, who was also active in developing electrical standards. He served with several prominent scientists of his time, including Clerk Maxwell and Lord Rayleigh. A book collector all his life, Clark tried to acquire all publications of value on the subjects of electricity and magnetism, as well as their applications. After his death in 1898 his collection of some 7,000 items, which included many rare books, was put up for sale. The British society most interested in Clark's work, the Institution of Electrical Engineers, noted in its journal in 1899 that "his library, so far as electrical works are concerned, was unequalled" and that there were few, if any, works of importance missing.[10]

William D. Weaver, who had written Carnegie in 1895 about a joint library, learned of the availability of the Clark collection and promptly informed a fellow electrical engineer named Schuyler Skaats Wheeler, who became interested and enlisted Carnegie's financial aid. Because negotiations were going slowly, Wheeler feared that the collection might be sold elsewhere and finally bought it with his own money, intending to give it to the AIEE. This move was admired by Carnegie, who then agreed to provide approximately $7,000 (a sum equal to the amount Wheeler had paid for the collection) for the shelving, indexing, and purchase of missing parts for the newly acquired collection. In May 1901 Wheeler officially gave the material to the AIEE. There were several provisos; any lack of compliance with them would cause the collection to be returned to Wheeler, his heirs, or his assigns. The conditions of the gift were that the collection be adequately insured against fire; that it be annotated and cataloged, with a copy of the catalog going to each AIEE member; that the collection be kept in New York City, available for use by the general public;

and that it be housed within five years in a suitable permanent home that was to be "reasonably safe from fire and not heavily mortgaged."[11]

The AIEE agreed to these conditions in 1901. This gift was to prompt efforts by AIEE members to seek funds for the building called for in the Wheeler gift. Later in 1901 a trio of AIEE members went to visit Carnegie to solicit funds for an AIEE building. Carnegie was pleased with the idea and agreed to contribute to a building fund, but he did not like the plans in one regard—he objected to the inclusion of a restaurant and dining room. He invited the trio to return with plans eliminating that feature, and they agreed. However, as evidence of his sincerity he announced his gift of $7,000 for indexing the Wheeler collection. Word of the meeting spread among members of other engineering societies. If Carnegie were willing to contribute to a building for the AIEE, they reasoned, he might be interested in plans for a larger building for the four major societies. Representatives of several societies had dinner with Carnegie when he was in Pittsburgh and urged him to think in terms of a joint headquarters. Carnegie encouraged this concept.[12] However, no definite progress was made at that time, despite Carnegie's enthusiasm. Later in 1901 the well-known electrical engineer Charles P. Steinmetz, president of the AIEE, also wrote Carnegie about funds for an AIEE building, but the response was not favorable.[13] Perhaps the talks about a building for several societies dampened Carnegie's enthusiasm for a building for just one society.

The Wheeler gift was to figure more directly in cooperative efforts when the AIEE building committee, working to find funds for the permanent home called for by the Wheeler deed, invited AIEE President Charles F. Scott to its October 1902 meeting. Scott had been one of the attendees at the dinner with Carnegie in Pittsburgh in 1901 at which a joint society building had been discussed. Before Scott attended the meeting with the building committee he asked William M. McFarland, an officer in Scott's company, for ideas on the design of a building suitable for four societies, then sketched a plan quickly. McFarland assured Scott it was feasible and encouraged him to carry it out, even though the plan embodied the rather unwieldly feature of four separate entrances for the four different societies. At the meeting he was dismayed that the committee was thinking on quite a modest level—a building for the AIEE alone to cost $50,000 to $200,000. When Scott described his ideas for a four-society building, he wrote later, "my air castle was greeted with silence." The committee members said it was theoretically all right but not practical. Scott continued, "I said that if it was all right theoretically it was good enough for me to talk about. So the meeting adjourned without our choosing plans or finding funds."[14]

Scott found another opportunity to discuss his "ideal" later that year when he was invited to speak at the twenty-fifth anniversary of the Engineers' Club of Philadelphia. He called cooperation the great discovery of the century and urged that engineering societies should join in creating a fine "Capitol of Engineering" as a sign of cooperation.[15]

The issue was again raised publicly on 9 February 1903, at the gala annual AIEE dinner held at Sherry's restaurant in New York City. The dinner was planned around the theme of celebrating the founding of the AIEE library. The dinner chairman, T. C. Martin, had finally been successful (after seven invitations) in persuading Carnegie to attend, along with several other well-known guests. Evidence shows that the real purpose of the library theme of the dinner was to bring the need for a joint engineering building to Carnegie's attention again.[16] It was a happy occasion, as Scott remembered it, with Carnegie in a good mood. Scott used the dinner meeting as an opportunity to renew his plea for a jointly supported building for the societies, and Martin aided Scott by saying, "I do not believe that anything would give Mr. Carnegie greater pleasure than to see our sister societies do as we have done and join hands with us so that our libraries might be of benefit to the world...."[17]

One of the speakers that evening was R.R. Bowker, who, as a publisher of serials and monographs of interest to both librarians and publishers, was a strong force in the United States for cooperative projects. In his talk on "the unity of science," he gave support to the idea of a central library for engineers; remarking "The generations past built cathedrals; the library is the cathedral of today. And no nobler cathedral could be made than one in which the representatives of the several engineering sciences should meet together with the equipment of a great library which would answer all the questions which could possibly be asked of the past." Another speaker was the director of the New York Public Library, John Shaw Billings, who stated that he saw no objection to having a strong engineering library created in New York City. It would serve a different clientele, on the whole, than his library would, and "the special technical library of the special society, if so managed as thoroughly to enlist the interest of the men who use it, will receive many things which would not ordinarily come to the public library."[18] Carnegie also spoke briefly to the assembled group of nearly two hundred members and their guests. He recalled some of his experiences as a youthful telegrapher, an appropriate topic for an evening sponsored by electrical engineers.[19]

Evidently Carnegie was impressed by the speakers' remarks about a joint library, for the next day he invited Scott and Calvin W. Rice (an old friend of Carnegie) to visit with him in his home. As Scott remembered it, Carnegie was interested in their desire to create a joint engineering building, although they had no specific plans ready at that time. "As we sat together on the big sofa, he between us, he said he thought that we 3 Pittsburghers might talk it over together. (Rice then lived in the 'Smoky City') and so we did....At the end of the hour, he bid us counsel quietly with a few others and come back when our plans were worked out, and he politely intimated twice more that we had better not come again until we had the plans."[20]

So strong was Carnegie's influence that the next day (Wednesday) eleven unofficial representatives of the various societies and of the Engineers' Club met at the latter's headquarters at 375 Fifth Avenue and discussed the situation. It was the consensus of the group that the time had finally come for a cooperative step. On the following Saturday six of the eleven representatives called upon Carnegie, who found them waiting for him at his home upon his return from the golf course. After greeting them, he expressed his surprise, saying that he had not expected them back until their plans were ready. Scott recalled that they then said: "we have them" and showed him a single typewritten sheet, which contained a brief analysis of the estimated cost of a joint engineering building. "I recall only the last line '$1,200,000'. There was discussion and explanation and then a lull. Presently he asked 'What is to prevent closing this up now?' There being no objection he went to his writing desk and with the ease with which he might have penned a letter of introduction he wrote and read to us with evident pleasure our $1,000,000 valentine (it was February 14, 1903)." The body of the note, addressed to the four societies and to the Engineers' Club, was as follows: "It will give me great pleasure to give, say, one million dollars to erect a suitable Union Building for you all, as the same might be needed."[21] Scott recalled, "When he [Carnegie] came to the words 'Say, one million dollars' his eye twinkled and he moved his hand up and down adding 'more or less'."[22]

With this simple note Carnegie set in motion a plan which, though requiring several years to come to fruition, would provide not only a fine headquarters and library building, but would also help serve to unify the major engineering societies in a way never before possible.

CONSTRUCTION OF THE BUILDING

One of the first problems was to select suitable property. The Engineers' Club had purchased a lot in February 1903 on West Fortieth Street, across the street from the new central building being constructed for the New York Public Library.[23] It was subsequently decided to aim for lots 23-33 on West Thirty-ninth Street, which abutted the club's lot. The lots for the societies' building cost about $540,000.[24]

With his usual clever business instinct, Carnegie counseled against a public announcement until the options had been safely obtained on all property needed for the two buildings, to avoid the inevitable escalation of prices that would occur if the project became common knowledge. The announcement was finally made on 30 April 1903, at the Engineers' Club. There was general acclaim of the plan. In an article printed a few days later reporting on a meeting of representatives of interested parties, the *New York Times* stated, "His [Carnegie's] object, and that of the engineers who enlisted his aid, is to make New York a sort of international engineering centre, at least of the Western Hemisphere."[25] The account went on to name John Fritz, whom it described as "the aged ironmaster of Bethlehem, Pennsylvania, an intimate friend of Mr. Carnegie," as one of the major leaders in securing Carnegie's aid. Fritz had indeed been a friend of Carnegie for a long period, as their acquaintance began when both were working in the steel industry, hence his urging of Carnegie to aid the building project was undoubtedly a significant factor in Carnegie's willingness to participate. Carnegie was quoted as saying that the purpose of his gift was to show the world the spirit of cooperation among American engineers in contrast to the more independent nature of European engineering societies.[26]

Within a short time the proposed joint building was voted upon and accepted by the members of three of the four societies— the ASME, the AIME, and the AIEE. At one point in the period prior to the AIME balloting Charles Scott, one of the prime movers in the proposed building project, felt that the AIME secretary, R. W. Raymond, was opposed to the project, partly because of a fear that the unified library to be created would not be able to equal the quality of service given by the AIME library. Upon being reassured by Scott that the several societies would each have a hand in determining the extent and type of library service offered, Raymond then swing his support to the project.[27]

Then a problem arose early in March 1904 when the ASCE announced that a poll of its members on the proposed project resulted in a vote of 1,139 to 662 against the plan.[28] No doubt purchase of the ASCE headquarters building on West Fifty-seventh Street as recently as 1897 was a factor in the negaive vote.[29] As a result of the setback, a subcommittee of the Carnegie Conference Committee visited Carnegie's home a few days later to learn his reaction. He sought reassurance tha the other societies would not reject a renewed offer of money.[30] AIEE's Scott described the meeting with Carnegie: "With evident disappointment equal to our own he said 'Then we can't go ahead, can we?' Then we presented letters from many smaller engineering and scientific societies asking for quarters in the proposed building. 'Then they can take the place of the "civils" and we can go ahead.' His smile returned." So on 14 March he wrote a new note addressed to the three interested societies and the Engineers' Club, increasing the amount of the gift to $1,500,000.[31] As can be seen from a reproduction of this document, it was written in the same informal style of his first offer the year before.[32]

ANDREW CARNEGIE
2 EAST 91ST STREET
NEW YORK

March 14th – 1904

Gentlemen of

The Mechanical Engineers
Institute of Mining Engineers
Institute Electrical Engineers
Engineers Club New York City

It will give one great pleasure
to devote say one (na) half million
of dollars to erect a Suitable Union
Home for you all in New York City

With best wishes
Yours truly
Andrew Carnegie

From Carnegie Papers, Library of Congress.

Fortunately the rejection by the ASCE of the merger plan did not halt or even particularly slow it down. Just four days after Carnegie's letter announcing his offer the joint committee of fifteen members was renamed the Engineering Building Committee, under the chairmanship of the influential Charles Scott of AIEE. The committee's main problem was that of deciding how to split the Carnegie money among the three engineering societies and the Engineers' Club. The decision was to allot $1,050,000 for the engineering societies' building and $450,000 for the structure for the Engineers' Club.[33] It was planned to make the engineering societies' building twelve stories high, with office space for each society and several auditoriums, one of which was to be large enough to hold 1,500 people. There was to be a "noble library hall" which would accommodate the 50,000 volumes held at that time by the libraries of the three societies, as well as the titles to be added in the future. The sponsoring societies together had more than nine thousand members and a total annual income of $135,000. Other societies having a membership of over 1,000 members were said to be desirous of being included in the building.[34]

In order to obtain the best possible building plans for the engineering societies' building, it was decided to have an open competition among architects in 1904. One million dollars was available for the building and half a million for decorations and equipment. Carnegie stated that the building for the societies was to be "stately, with no frills, so that its style will be as good a hundred years from now as it will be when it is new." The club's building was to be more ornate.[35] The holder of Columbia University's chair of architecture, Prof. W. R. Ware, was one of the judges of the competition for the winning drawings. The award went to Herbert D. Hale of Boston, grandson of the venerable author and orator Edward Everett Hale, and his associate Henry G. Morse of New York City.[36]

A major legal step was taken on 11 May 1904 when the New York State Legislature enacted a law creating the United Engineering Society, having the following goals: "The objects of the corporation hereby created shall be the advancement of the engineering arts and sciences in all their branches and to maintain a free public engineering library."[37] It is significant that the inclusion of a library was featured so prominently in the wording of the bill. The act furnished a legal way for the societies to hold property jointly.

During the next fourteen months events moved rather slowly, but progress was nevertheless made. For example, the engineers raised the half-million dollars to repay Carnegie for his loan of money for the cost of the property, and during the spring and summer of 1904 the old residences on the building site were removed. Little public news was made until the release of the building plans in May 1905. They called for a fifteen-story building, with thirteen of the floors above the ground (one more than the first description publicly released in earlier years). It was to have a frontage of 115 feet on Thirty-ninth Street, designed in what was called the French Renaissance style.[38] The following July the contract was let for the erection of the building to a New York firm, Wells Brother and Company.[39]

Work continued on the building until 8 May 1906, when there was a momentary halt for the laying of the cornerstone in a public ceremony. On hand were Carnegie, his wife, and their daugther Margaret. After the photographers had finished taking pictures, the brief ceremony began. The opening address was given by Eben E. Olcott, then president of the United Engineering Society, who thanked Carnegie for the gift and made note of the assistance the building would give the engineering profession. Then Mrs. Carnegie, with the aid of T. C. Martin, friend of the family and editor of *Electrical World*, placed in the cornerstone a box containing a set of newly minted U.S. coins, miscellaneous records, a Bible, and selected publications of the sponsoring societies. Also included was a gold-plated copy of Carnegie's letter of 14 March 1904. Carnegie then gave a brief impromptu speech:

> They have asked a word from me and I'm going to respond. I congratulate you on the near completion of this building. Here engineers can consult with one another and cultivate friendship. They can form a brotherhood, which will be a great benefit to all. Oh, we in America are great mixers. In Europe you will find a group of mechanical engineers in one place, electrical engineers in another, and mining engineers somewhere

else, all widely separated. Then they expect to keep
pace with this country, when we have a home like this.
The principle of union in Science is as important as it
is in politics.[40]

Although the building had been scheduled for occupancy by
1 January 1907, it was erected faster than anticipated. The UES
Board of Trustees met in the building for the first time on 22
November 1906, and by 1 December the structure was actually
completed, with the UES assuming management of it on 15 December.[41] The building rose high above a row of five-story
residences lining the north side of Thirty-ninth Street. At the
extreme right of the building was an entrance to a driveway which
curved around the structure, emerging on the left side. It was
designed for the use of carriages leaving the street to deposit
passengers at a special carriage entrance on the east side of the
lobby floor. (The eight-foot-wide drive became too narrow for use
when automobiles became more popular than carriages.) The
main entrance was centered on the front of the building, with
another vestibule at the extreme left side that led directly to the
elevators and stairs linking the floors. If for no other reason than
its relative height compared to the homes alongside, the building
was striking in appearance, with a rather ornate combination of
columns and high-arched windows at the third and fourth floor
levels, which were devoted to the building's largest auditorium.
Over the main entrance was inscribed "Engineering Societies."
In the ornate lobby were twelve columns of Swiss Ciptolino marble
and gold ornament on which emblems of the sponsoring societies
were mounted.[42] The society offices were located on the fifth
through the eleventh floors; the library stacks were on the twelfth
floor and the reading room on the thirteenth level.

The dedication of the building was the occasion for a gala
series of events, extending over the period 15-20 April 1907. On
the first day, Monday, there were technical papers read by various
engineers. The main event of the week took place the next day,
which featured noted speakers and the readings of greetings from
all over the world. Among the honored guests were Alexander
Graham Bell, Andrew Carnegie, Edward Everett Hale (whose

53

Laying the cornerstone for the Engineering Societies Building, 8 May 1906. Andrew Carnegie's daughter is in front, holding flowers; her smiling father stands behind her, and Mrs. Carnegie is to his left.

grandson was the building's architect), and heads of major engineering societies, both American and foreign.

A congratulatory message from President Theodore Roosevelt was read by T. C. Martin, then president of the Engineers' Club. It read in part:

> The building will be the largest engineering centre in the world. It is, indeed, the first of its kind, and its erection in New York serves to mark and emphasize the supremacy which this country is steadily achieving through her proficiency in applied science. The whole country is interested in the erection of such a building, and particularly, of course, all of those who follow the profession of engineering or any kindred profession, and in no branch of work have Americans shown to greater advantage what we like to think of as the typically American characteristics.[43]

One of the speakers was Yale University's president, Arthur Hadley, who spoke of the need for engineers to have two quite distinct qualities in order to provide the best professional service—one was technical competence and the second, "for want of a better word, may be called the ethical standard." Charles Scott, in his introduction of Carnegie, said:

> You provided money. It has been transformed into brick and mortar. But your letter gives more than money: in it is an ideal: 'A union building for you all.' That, too, has transforming power. Already the harmonizing feature is an active force. Our societies are working together. They are getting a broader view of their position and a new inspiration for the large work that lies before them.[44]

Carnegie, after a few bantering words, urged his listeners to have faith in the ultimate chice of humanity for that which is right, saying, "As sure as the sunflower turns toward the sun so the human race turns toward better things.... So I look forward to the future of this building, and I know that the organizations to whom it is devoted will advance and continue to meet the develop-

The Engineering Societies Building.

ing needs of our age." He concluded with a reference emphasizing his long-term interest in world peace: "We have only to go on to know our brothers the world over in order to realize that all men of all nations are indeed brothers."[45]

Later that evening there was a round of receptions in the new building, the first open to all guests in the main auditorium area, followed by smaller gatherings held by the various societies in their quarters. Towards the end of the week the John Fritz Medal was given to Alexander Graham Bell for "the invention and introduction of the telephone." This was followed by the presentation of medals to Rossiter W. Raymond, Frederick R. Hutton, and Ralph W. Pope in recognition of their long years of service as secretaries of the mining, mechanical, and electrical engineering societies, respectively; each had served twenty-five years or more. Many technical papers were presented up through Thursday of that week. On Friday, the 19th, a vaudeville show was presented in Madison Square Garden for invited guests. On the next day there were scheduled visits to several of the major engineering projects underway in the city, such as subway tunnels. It appears that the dedication committee, under the chairmanship of AIEE's John W. Lieb, Jr., had done a commendable job of heralding the opening of the new building.[46]

THE NEW LIBRARY FACILITIES

Thus after many years of dreaming, talking, planning, organizing, and fund raising, the joint home for the societies was finally officially dedicated. Actually it had been declared ready for occupancy as early as December 1906, and the three societies were free to set their own moving schedules. All did so early in 1907. Their libraries were installed on the twelfth floor (the stack area); their staffs, key periodicals, and reference materials were located on the thirteenth floor (the reference and reading room area).

As can be seen from the floor plan, the reading room floor was built so that the shelving formed eight alcoves, four on each side of the center row of tables. For several years each of the three

57

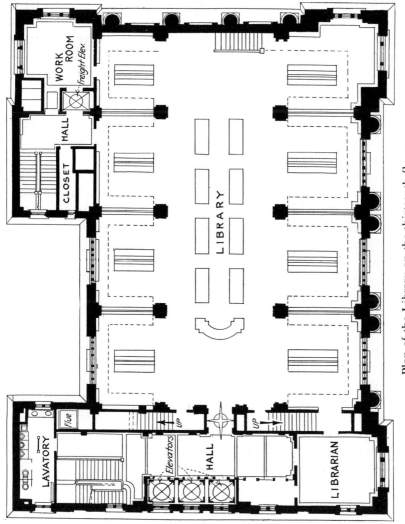

Plan of the Library on the thirteenth floor.

societies was allotted two alcoves (one on each side of the tables in the center), where they kept reference materials and bound volumes of their key periodicals. The remainder of their collections was shelved in the stacks. Since there were only three societies involved at that time, the fourth set of alcoves was devoted to general materials presumably of interest to all. Later changes in the reading room included adding small staircases leading to narrow balconies lining the alcoves, so as to provide acess to the upper tier of bookcases in the high-ceilinged room. At the far end of the room was a large mural painted by Frank Dana Marsh. As an article written a few years after the library opened indicates, the mural depicts "the operations of engineering. The central figure is that of the directing engineer, robust and keen-eyed...."[47] (The official name of the Library until 1918 was the Library of the Engineering Societies, but it is apparent from the title of the article that its present name, the Engineering Societies Library, must have been in common use long before the official change took place.) A floor plan in the article showed that by 1914 the tables extending down the center of the room had been removed, with small shelves for current periodicals replacing them. Skylights were visible over the card catalog and reference desk areas. The polished wooden floors, decorative figures on the columns at the end of each alcove, and the graceful ceiling treatment made the room attractive and dignified.

While the layout afforded a pleasant-looking reading room, it provided little space for the staff. The librarian was given a spacious office at the Thirty-ninth Street side of the floor, but there was only a small workroom for the technical services staff. The reference staff worked in the open center of the room, with no offices at all. Lacking definite evidence, one can only speculate about the amount of consultation the architect had with librarians before preparing his design of the library.

Early descriptions of the building indicate that the purpose of placing the Library on the two top floors of the structure was to increase the amount of light, to reduce noise and dust, and to obtain "freedom from flies."[48] Charles Scott wrote enthusiastically about the view from the Library: "From this eminence [the thirteenth floor] one could see both rivers; looking south the first

The Library reading room.

high building was the Waldorf-Astoria Hotel of about 14 floors (where is now the Empire State Building) while to the north the only high standing edifice was the Times tower."[49]

Former employees who worked in the building remember many pleasant features of the Library, particularly the charm of the reading room. But they also remember that there were problems with the building, such as the fact that dirt accumulated readily or that the tall windows had only small panels which could be opened for ventilation, making them ineffective on hot days.

A brochure published by the UES as the building was nearing completion was designed to interest prospective tenants. The Library received particular attention as an important feature; as the brochure stated, "The Trustees hope that all organizations who are to be welcomed to the building will add their libraries to the other units, under proper arrangements as respects proprietary rights and titles, so that this library may become one of the notable features of the building. It is to be conducted as a free public library of reference."[50]

5

Notable Library Personages, 1907-46

It is probably a truism that the characteristics of an organization are largely reflections of the objectives and attitudes of the persons responsible for its establishment and operation. The Engineering Societies Library is no exception to this rule. Over the years its nature has been shaped by the outlook and goals of many individuals. The aim of this chapter and of the one which follows it is to review briefly the contributions of those who have played a major role in the development of the Library, including its directors (a surprisingly small group) and selected employees and board members. Space limitations do not permit citing the role played by many others, including several outstanding employees and those who served many hours on Library committees and on its board.

LIBRARY DIRECTORS

Probably the most indelible marks made on the nature of the Engineering Societies Library were due to the handful of people who have served as its director in the years since its founding in 1913. There were, of course, outstanding individuals whose vision and determination led to the creation of the Library, and the previous chapter gives the highlights of their contributions. This section begins in 1907, when the leaders of the three engineering societies (representing the disciplines of mechanical, mining,

and electrical engineering) had successfully worked with Andrew
Carnegie and their own constituencies to bring into existence
the new Engineering Societies Building, into which their individual libraries moved during the early months of 1907. For the
remainder of that year and into the summer of 1908 thee was no
single person in charge. Rather it was a case of each of the three
libraries retaining its own staff, operating as a loose confederation
of libraries. The United Engineering Society took the much-
needed step of appointing a chief librarian in the summer of
1908.

L. Elizabeth Howard (1908-10)

Howard was given the authority to supervise all employees
of the Library, a decision made by the Library Committee, as the
governing body over the Library was called at that time, within a
year of the move into the new building.[1] They saw the impracti-
cality of retaining three separate staffs and the need for unification
of effort. Howard had served for several years as head of the library
of the American Institute of Mining Engineers (AIME), which
had been concerned with useful but routine sevices, although not
as innovative as some of the services contemplated about that time
by the ASME library, such as fee-based searches.[2] She was in
charge of AIME library when it moved into the new building
in 1907 and thus was aware of the background of the situation, as
well as being experienced in operating an engineering library. In
1909 she brought in Alice Jane Gates to serve as assistant
librarian, a post she was to hold until 1917.

For reasons that have never been determined, Howard resigned
as chief librarian in August 1910, leaving Gates as acting head of
the Library until a new chief librarian was named.[3] Howard had
such a short tenure as head of the Library that she hardly had
sufficient time to make much progress in improving conditions,
such as finding the best way to deal with the three separate library
staffs and collections. From studies published later it is known
that the total collection size in 1910 was less than 45,000 volumes.
Each of the libraries had its own cataloging system, and nothing
was being done to integrate the three catalogs. Difficult problems

faced the next incumbent, William Cutter, who was appointed to the position of chief librarian before the end of 1910 and took office in February 1911.

William Parker Cutter (1911-17)

William Cutter was a nephew of Charles Ammi Cutter, the prominent librarian known for many accomplishments, especially his outstanding monograph on the construction of a dictionary catalog and his tables for the assigning of author numbers to books. It is clear, however, from William Cutter's gracefully written little biography of his uncle that the two were not very close: "When I reach a description of the man [Charles] Cutter himself, not the librarian or the author or the camper, but the inner soul, I must rely largely on others. I never had personally the intimate acquaintance with my uncle that was enjoyed by those more closely associated with him. Quotation must be relied on, therefore."[4] William Cutter at least had a part in bringing to publication a posthumous publication of his uncle when he edited the last edition of Charles's well-known monograph on dictionary catalogs.[5]

William Cutter, at the age of forty-three, came well prepared for the job at the UES Library. He was graduated from Cornell University in 1888 with a degree in chemistry, then spent five years as a chemist at the agricultural experiment station of Cornell and Utah.[6] He began his career as a librarian as head of the library at the U.S. Department of Agriculture from 1893 to 1900; he was the first one holding that position to have been appointed under civil service rules. He reorganized the library, built up the staff with trained people, and introduced modern methods. He was said to have been particularly successful in obtaining adequate annual appropriations for the library and in establishing two publications, a library bulletin and a bibliography series. Under Cutter the library became the first one to make available printed catalog cards for its own publications, and he also established branch libraries, the first being at the Bureau of Animal Industry. In January 1901 he became chief of the Order Division of the Library of Congress. Here he became interested in rare books and contributed a chapter on the subject to a work on book collecting.[7],[8]

In 1904 Cutter succeeded his uncle as librarian at the Forbes Library in Northampton, Massachusetts;[9] this was the same library in which Alice Jane Gates had been employed from 1906 until she came to the Library of Engineering Societies in 1909. By 1911 William Cutter saw the collection there grow to nearly 115,000 volumes, larger than the average for towns the size of Northampton.[10]

Cutter's acceptance of the job at the Library of the Engineering Societies in 1910 may have reflected a desire to return to a technical library, one that was becoming nationally known and had great potentiality. It was not until February 1911 that he actually began his new job, where he found three separate libraries—one for each of the Founder Societies. The confederation of libraries was still governed by the three-man committee, made up of the heads of the library committees maintained by each of the three Founder Societies.

However, in 1912 a recommendation of the Library Committee was to lead to the beginning of a library that was truly a merger of its constituent parts and that was under the control of the United Engineering Society, not governed by a committee whose members were responsible only to their individual societies. It is not known exactly why the Library Committee recommended that the Library be put under the direction of a new governing body responsible to the UES; perhaps experienced backers like Charles Scott, who could see the handicaps under which the three libraries were operating, were able to convince Library Committee members of the need for a change of governance. At any rate, it was the proper decision to make if the Library were to succeed and serve its users well. There is at least a record of how the ASME Library Committee felt about the proposed changes; in general, it was enthusiastic. The chairman of that committee was, of course, a member of the three-man Library Committee that made the recommendations, and he acknowledged in the report how he had been fully aware of the great desire of the late Charles Wallace Hunt, former secretary of the ASME, that a better plan be found for governance of the Library and of his regret that Hunt died a year before the necessary steps were taken. The report stated: "Through the initiative of the Library Committee amendments to the By-Laws of the United Engineering Society have been proposed whereby a common administration of the joint library...may be secured, making possible greater efficiency in the joint library service, together with symmetry of the several collections, so that there may be no duplication of effort."[11]

By November 1912 each of the governing bodies of the three Founder Societies had ratified the recommendations, at which time new bylaws for the UES went into effect.[12] The new bylaws read in part:

74. The Board of Trustees shall maintain and conduct a Free Public Engineering Library, subject to such regulations as it may from time to time determine.

75. The library shall be conducted by a Library Board, subject to the direction of the Board of Trustees.

76. The Library Board shall consist of sixteen members, composed as follows:

Four members designated by each of the three Founder Societies.

The Secretary of each of the three Founder Societies.

The Librarian, who shall also be the Secretary of the Board.[13]

It was to be a sizable library board; it was based on the assumption that the busy secretaries of the societies would have the time to devote to this purpose. This board, however large and cumbersome it might be, was an important factor in making the joint library a viable organization. The new conditions for library operation were to be so different that the Engineering Societies Library (though not yet officially so named in that period) can justifiably be said to have been created in 1913, when the newly formed Library Board first met.

The first meeting of the board (on 6 February) was a historic one since it marked the end of the control of the Library by the library committees of three separate societies. For Cutter it marked the beginning of a period when the organizational problems of how the sponsoring societies could provide for a common library for them all had been largely solved. As secretary of the board he was ensured a role in the discussions and deliberations of the board, a position strengthened by his additional duty of serving as secretary of the board's Executive Committee.

Two items submitted by Cutter received no action at this first meeting: one was a communication about the Library's quarters (which was postponed for later consideration); the other was a proposed budget (which was referred to the Committee on Budget).[14] Although nothing momentous occurred, the important point is that the Library Board was functioning. It was responsible to the UES Board of Trustees, not to the individual societies, and all plans for funding had to be approved by the UES board.

The Executive Committee of the Library Board held its first meeting on 4 March 1913. At that meeting the board began consideration of the overcrowded shelving on the twelfth floor, a topic that reappeared on the agenda the following month, when a project to install more shelving was duly approved. At the April meeting the committee discussed a proposal for a downtown branch of the library; it was to be based on collections donated by the New York offices of the General Electric and Westinghouse companies, from whom the suggestion had come. An even more venturesome but evidently less realistic project was discussed at the June meeting; the chairman reported on a talk he had with John Hays Hammond, an engineer of note at that time, regarding the possibility of opening a branch of the library in San Francisco. There was no further reference to this topic. In contrast to the prospect of a California branch the committee also had mundane business to take care of—it approved the hiring of an additional library assistant with the pay set at $780 per year.[15]

Cutter's tenure of office started off well enough when the board showed its support for him and the library staff by granting a 10 percent pay increase for all employees effective in October of his first year, with his salary established at $3,900 for that year.[16] However, as time went on conditions worsened.

Cutter had many ideas abut improving the Library and its services, one of which concerned the disturbing fact that each of the three merged libraries had used its own form of the Dewey Decimal Classification system. He urged the Library Board to let him begin a project to completely reclassify the collection, a recommendation which the board ignored.[17] Another project he favored was the creation of an improved index to cover the world's technical periodicals on engineering topics; he stressed the need for inclusion of foreign titles, for a subject portion arranged by a classification scheme, and for author/subject indexes.[18] Again, his comments went generally unnoticed.[19],[20]

Another sign of lack of support for Cutter was found in the board's creation of a section of the library which was to provide reference and translation service for a fee. It was the sort of service that Cutter was eager to supervise, yet to his probable surprise the board set up the Library Service Bureau, as the newly formed unit was named, so that it reported directly to the Library Board's Library Service Committee, rather than to Cutter. It was not an altogether new service in the library (only the name and pattern of organization had changed), so there is no question about Cutter's ability to supervise such an operation. The board's handling of the matter was a slap in the face for Cutter.[21]

Cutter's term as librarian came to a close in 1917. It is surprising that he stayed as long as he did, when one reviews the many ways in which the Library Board's Executive Committee constantly denied him the level of authority normal for the position he held. This committee took much of his responsibility on itself, or gave him simpleminded tasks to do, spelled out as if he were a schoolboy needing detailed instructions to do commonplace assignments properly. For example, the committee took the following actions: asked Cutter "to purchase a ruled book for record of searches" made;[22] appointed a committee to decide whether to use copper plates or steel ones for preparing the engraved library stationery;[23] asked Cutter to submit lists of proposed books to the committee for approval, then get bids from dealers, then return to the committee for a final word of approval on what to buy;[24] established the Library Service Bureau, which was to report to the board's Library Service Committee, rather than to Cutter;[25] asked their House Committee and board chairman to find the best location for the new photostat machine;[26] conducted its own experiments with lighting for the reading room, had a special "bellshaped fixture" designed and asked the House Committee to supervise the installation.[27]

These examples indicate the relationship Cutter had with the Executive Committee, which seemed to have little or no faith in his ability or his judgment. Perhaps the board members thought he had a weak personality. Whatever the cause of the friction, Cutter endured the lack of support by the committee longer than many librarians would have. Finally at the 7 February 1917 meeting of the Executive Committee it was announced that Cutter had resigned. At that same meeting the committee made known that it was recommending to the UES Board of Trustees the appointment of Harrison W. Craver as Cutter's successor. The committee proposed a salary of $7,500 for Craver, with the title of director of the Library. It is not recorded how long the committee had known of Cutter's decision to resign. If it had been a matter of months, then one could understand how the committee had had the time to search for and decide upon Craver as Cutter's replacement so soon. Another conclusion that could be drawn is that the committee decided not to rehire Cutter and found a successor, then asked Cutter to resign and thus save face. This act of consideration, if it actually happened, was followed by another—that Cutter was to be paid for the remainder of 1917 unless he found another position.[28]

There is no record that Cutter was employed the remainder of the year, so perhaps he took advantage of the board's offer. There is, however, one indication of how he felt about the Library; in his biographical sketch in *Who's Who in Library Service* he did not even list the fact that he had been the chief librarian at the Library of the Engineering Societies.[29]

The way in which Cutter's talents were ignored by the board's Executive Committee is difficult to understand, in view of his earlier library posts. Also there is much evidence in his writings that Cutter showed great interest and pride in the Library. For example, one of his articles revealed his obvious conviction of the worth and skill of the reference service it was offering.[30] In another article he praised the collections and physical facilities of the Library.[31] He showed his keen interest in improving the quality of periodicals, the most important form of literature in most technical libraries.[32] Cutter was not a misfit as a librarian;

he was professionally well recognized through his writings and the posts he held. Why he never received proper recognition from the influential Executive Committee remains a mystery.

Cutter had become librarian at a crucial time for the Library. It was a period in which various organizations and financial arrangements for the operation and support of the Library were being evolved by the Founder Societies and the UES officials; three different library collections were to be merged; and preparations soon had to be made for the moving in of a fourth large collection. Through all his term he had to deal with certain board members who stripped him of many major responsibilities for no reason. At the time he left, or shortly thereafter, six other staff members resigned, including Alice Jane Gates, the assistant librarian.[33]

Cutter went on to become the head of the library and information department of well-known firm, Arthur D. Little, Inc., in Cambridge, Massachusetts (1922-28). Then he became assistant librarian of the Baker Library at Harvard University, where he prepared a special classification system for that noted business collection.[34] His last position was that of librarian for the Bermuda Biological Station for Research, at which post he died in 1935. He had been an active, professionally oriented librarian throughout his career.[35] His contribution to the Engineering Societies Library would have been much more significant he he had board support.

As a postscript to this summary of Cutter's career, it should be noted that a major event had taken place near the middle of his term of office. After several years outside the fold, the American Society of Civil Engineers (ASCE) took actions that were to lead to its becoming the fourth Founder Society. Informal conversations held in 1915 between officers of the United Engineering Society (UES) and certain ASCE officials resulted in an invitation being offered the ASCE to join the UES.[36] An estimate of the cost of adding two more floors to the UES building was approximately $250,000. As for operating expenses, the ASCE library cost about $7,000 per year versus only $4,000 as the ASCE annual share of the UES Library upkeep.[37]

ASCE members roundly supported the idea in June 1916, approving it by a lopsided vote of 2,500 to 390.[38] The addition to the building was completed later that year, and that winter the

ASCE library began moving into its new quarters as part of the UES library, completing the move in March 1917. This added some 67,000 volumes to the Library.[39]

With the addition of the ASCE, the total membership of the four societies making up the UES had grown to more than 28,000 by 1916. (The four had a total of 11,000 members in 1903.) The total cost of the land and the building, including the three-story addition, was approximately two million dollars; Carnegie had given $1,050,000 of this amount (he had also given $450,000 to the Engineers' Club). Cutter left at a time when the UES and its Library were beginning to grow in size and prominence.

Harrison Warwick Craver (1917-46)

The selection of Harrison W. Craver as William Cutter's successor ushered in a period during which the Engineering Societies Library achieved complete amalgamation of its original separate libraries into one unified whole, and began to receive greater national recognition as an important technical library. Craver was to serve nearly twenty-nine years, thus spanning from the final years of World War I to the period just after World War II. Many important social and economic events of that era were to leave their mark on the operation of the Library.

Craver's Background   Craver was born in Owaneco, Illinois, on 10 August 1875. He went to a college fairly well known in the Midwest, Rose Polytechnic Institute in Terre Haute, Indiana, from which he was graduated in 1895 with a bachelor's degree in chemistry. (The school awarded him an honorary doctorate in science in 1933) He was attracted to the burgeoning steel industry in Pittsburgh, where his first position was that of a chemist at Kirkpatrick and Company. In 1900 he became interested in technical literature and began work in the technology section of the Carnegie Library of Pittsburgh, where he became technology librarian in 1902. Later that year he decided to return to industry as superintendent of the Allegheny Iron and Steel Company, but before the end of the year he returned to the Carnegie Library to take charge of the Division of Technology, at that time one of the sections in the Reference Department. In 1903 he was promoted to be the first head of the new Department of Technology, which had grown so rapidly that it was made equal in status to the Reference Department. Years later, chatting with an employee about this period of his life, Craver said that the pressures of industry caused him to seek a quieter profession, yet one which involved working with technical literature. Some of his associates in the steel industry subsequently became millionaires but died before they were fifty years old because of the stress of the work.[40]

Craver became a respected member of the library staff, with full use made of his technical background. The library's annual report for 1903 stated that:

> The growth and importance of the work in connection with the reference collection of the literature of technology made it seem advisable during the year to raise what was formerly a division of our work [the Reference Department] to the rank of a regular department. The head of the department [Craver] is the expert advisor for the entire library system with reference to technological literature, and his services are invaluable.[41]

By 1908 Craver had increased the Technology Department's collection to 50,000 volumes (out of nearly 300,000 volumes in the whole library), and had done well enough to merit his appointment as librarian.[42]

As head of the library Craver found that his duties put him in contact with many affluent, influential people. He was described as having the knack of getting along well with people of importance, such as Andrew Mellon, said to be courteous but painfully shy, who served as a member of the Library Board and chaired its Auditing Committee.[43] Craver's array of prominent acquaintances was such that he found it natural to correspond directly with Andrew Carnegie about library matters. One of Craver's early letters as librarian was written to Carnegie, enclosing a copy of a two-volume book catalog for the entire library holdings updated as of 1906.[44]

In 1910 it was decided, perhaps by Craver but certainly with his approval, to construct a classed card catalog for the Department of Technology. The book catalog published in 1908 was

based on shelflist cards arranged by call number in order to produce a classified catalog, but the main card catalog for the library
was in dictionary form. The choice of a classified catalog was
based on the opinion "that for technical works the classified form
would be more satisfactory than the dictionary form....As the
technical books are catalogued for the classified catalogue and
also for the general dictionary catalogues, a good opportunity
is given to compare the merits of the two forms."[45] Craver was to
carry this interest in a classified catalog with him to his next
position.

In the winter of 1917 Craver was offered and accepted the
position of director of the Library of the Engineering Societies
in New York. His letter of resignation from the Carnegie Library,
written on 23 February 1917, asked that the effective date be 31
March which was accepted by the Board of Trustees at its 5 March
meeting. In the letter Craver mentioned his regret at having to
sever "pleasant relations" and also expressed appreciation for
the "loyal, intelligent and friendly associates" he had at the
position he was leaving. In reply the board commended him for
the "extremely valuable work which he has done in expanding
the usefulness of the Library and in making it a power for good
in the city of Pittsburgh."[46]

Early Years of Craver's Term   Craver was appointed director
of the Library of the Engineering Societies sometime between the
February and March meetings of the Executive Committee of the
Library Board; he was present for the first time at a meeting of the
Executive Committee on 4 April 1917, four days after his resignation from the Carnegie Library had taken effect. In accordance
with the bylaws governing his new library, he became secretary
of both the Executive Committee and the Library Board.[47]

The board and the Executive Committee soon began to grant
him authority in regard to matters that they had previously
handled themselves or had assigned to committees rather than
allot to Cutter. For example, at the May meeting he was given the
authority to decide whether or not to retain two separate staffs to
handle reference work, one group for service done at no charge,
and the other group—the Library Service Bureau—for research
done for a fee. This was the organizational arrangement the

Executive Committee had established in 1915; Craver retained
the two separate staffs. A few months later he was empowered to
increase service charges for the work of the Library Service
Bureau as he felt appropriate.[48] This early assignment of responsibility to Craver indicated the confidence the board had in him.

Several personnel matters received attention by the board and
the Executive Committee early in Craver's term. One was a decision
made in May 1917 to grant all employees a three-week vacation,
a move that put the staff more on a par with employees of many
public libraries.[49] In September members of the Executive Committee found on the agenda a letter from the president of the
United Engineering Society in which he urged the committee to
cancel the agreement regarding Cutter's salary. With the approval
of the UES trustees, Cutter had been granted full pay till the end
of 1917 if he did not take another position during that year. The
records do not indicate what prompted the president's letter; at
any rate, the Executive Committee stood firm and refused his
request to cancel the arrangement. Evidently Cutter was under
fire even after he left the Library. Another staffing item was that
of resignations—three staff members resigned within a month of
Craver's arrival, followed by three more by the following September. No reasons were recorded in the minutes for these departures,
although the arrival of a new head of an organization is often the
occasion for staff changes of this sort.[50]

Craver was on hand for a celebration held in December 1917
to honor formally the ASCE's joining of the United Engineering
Society. The day's events included a luncheon for ASCE officials,
and several talks that evening in the auditorium of the Engineering Societies Building, to which "ladies" were invited. (It is most
unlikely any women were members of the four Founder Societies
in those days.) Andrew Carnegie expressed pleasure "to hear that
the Civil Engineers have accepted the invitation to live in the
home of the other engineers."[51] The new quarters of the ASCE
were open for inspection afterwards.

Early in 1918 the question of a new name for the Library
was discussed by the board. At its January meeting, in consideration of a recommendation of the UES House Committee, the
board recommended to the UES Board that the library name be

changed to the "Engineering Societies Library"—a shorter, less cumbersome title than "Library of the Engineering Societies."[52] This change was subsequently approved by the UES, and the new name was officially announced at the February meeting of the Executive Committee.[53] It should be noted that on numerous prior occasions Cutter, for example, had listed himself in some of his periodical articles as working at the "Engineering Societies Library," so the new name was not entirely unheard of in library circles.

Finances were a constant concern of the board, and in 1924 it began to give serious consideration to seeking donations from industrial firms, a search for funds that was to continue throughout succeeding years. A boost to library funding came in 1926 with the gift of $100,000 by Edward Dean Adams, a member of the AIEE who had been a board member since 1915; the money was given to the UES with the proviso that half the income from the principal was to be used for special projects in the Library, not for regular operating expenses. Craver's personal finances were also improved in 1926 when his annual salary was increased to $10,000.[54] A few years later the endowment funds of the Library benefited from a gift of $50,000 by James H. McGraw, of the McGraw-Hill Book Company.[55]

While the various gifts of money were important to the Library, its main source of funds for many years was the payments from the sponsoring societies. At the time of the establishment of the Library in 1913 the funding formula called for each society to pay a lump sum of equal amount, originally about $5,000 per year. In 1925 it was estimated that this method of support cost each society an average of sixty cents per member per year.[56] By 1928, however, a more equitable plan was sought in order to make allowances for the differences in size of the various societies, and it was decided that each society would pay annually $5,000 plus thirty cents per member. The plan was in force during the remainder of Craver's term.[57]

For reasons not known at this time, the United Engineering Society changed its name twice in the early 1930s. On 1 January 1930 it became the Engineering Foundation, Inc., only to change that name in 1931 to the United Engineering Trustees, Inc.; it has retained the latter name since then. One of the societies also changed its name during Craver's term; in 1919 the American Institute of Mining Engineers became the American Institute of Mining and Metallurgical Engineers, with its initials remaining AIME as before the name change.[58]

Since the early years of Craver's term of office started out smoothly as far as his relationship with the Library Board was concerned, he was soon able to institute changes needing the board's approval that Cutter had sought but was rarely able to obtain. Whatever the reason for Craver's success with the board, whether it was his manner of doing business or his excellent credentials for the position, he made the most of his opportunities. For example, whether it was due to his efforts or not the Library's annual report for 1917 (his first year there) reported that publishers had begun to give the library free copies of technical books in exchange for book reviews prepared by the staff for use in engineering journals published by Founder Societies. This source of books was to grow in importance over the years, amounting to a significant cumulative sum.[59]

One of Craver's few failures in dealing with the board was his lack of success in persuading it to purchase *Engineering Index*, which had fallen on hard times and was available for sale in the fall of 1917.[60] But he never gave up, as one of his deepest interests was to involve the Library in the preparation of a strong index to engineering literature. Before the next year has passed he was probably the one who persuaded the Library Board to allow him to begin an index on mining and closely related materials, using the interest from the $100,000 gift made by James Douglas.[61] By the end of 1918 he had begun indexing journals on mining and metallurgy, making the most of a request from the Canadian Mining Institute for such a service to obtain board approval. Although there is little more said about the role of the Canadians, within a few years Craver was able to interest the AIME in having the Library prepare an index on these topics.[62] In a related matter he took advantage of the purchase of *Engineering Index* by the ASME in 1919 by instituting ways in which the Library could cooperate with the new owners of the *Index*, such as allowing the *Index* staff members to use library journals and books for indexing, thus saving them large sums of money. Craver had numerous ideas for indexing services, such as preparing annotations on card stock to send to subscribers to the service.[63] Although the board failed to go along with this plan, it is interesting to note that *Engineering Index* adopted the idea a few years later.[64] By 1928 the relationships between the *Index* and the Library were so close that the Library Board was invited to name a person to serve on the Advisory Board of the *Index*; Craver was appointed the alternate representative.[65]

James McGraw's 1929 gift to the Library allowed Craver funds for commencing a more elaborate periodical index, even though it meant competing in coverage with *Engineering Index*.[66] There is little known as to how much the Library index aided its staff, so that the wisdom of his pursuit of this goal can be questioned. However, one has to admire Craver's determination and his ingenuity in getting the various indexing projects approved and funded.

Another major interest of Craver's was classified catalogs, which he brought to the attention of the board soon after his arrival, pointing out the urgent need to bring all the books into one classification system.[67] The board granted him the permission to begin this project, which they had denied Cutter when he made a similar proposal. Although it was to be almost two years before the final approval was granted and funds allocated, Craver thereafter instituted quick action, persuading a former employee of his at the Carnegie Library, Margaret Mann, to serve as the head of the catalog department and also to direct the reclassification project. She did a commendable job and remained until the work was essentially completed before taking a post in Europe in 1924.[68]

Craver watched closely over the public service units of the Library and instituted the keeping of more accurate records so as to have statistics on this vital part of the work. For example, in 1918 he saw to it that the number of telephone requests were recorded. During the early years of his term the board agreed to allow the Library to make loans of books to members of engineering societies, although for many years loans were restricted to duplicate copies only.[69] One of the members of the public service staff during Craver's term of office was Ralph Shaw, later to become one of the country's best known librarians.

Later Years of Craver's Term   During World War II Craver guided the Library through troublesome times, making it invaluable to civilian and military users in spite of staff problems and difficulties with the receipt of important materials. In 1945 the U.S. War Department sent the Library a letter which praised the staff for its contribution to the war effort.[70]

Craver's devotion to the Library was evident throughout his term, yet he still found time to engage in many professional activities, including writing, giving papers at meetings, and being active in various library associations. He also had a scholarly bent, translating foreign technical articles or papers. Another example of his scholarship was the analysis he and two colleagues made of the diary of a nineteenth century illustrator, H. Farey, which was filled with tiny, meticulously drawn pictures of engines. It was not a work of any great important historically, but it illustrated Craver's interest in certain esoteric subjects.[71]

In 1937 the American Documentation Institute (now the American Society for Information Science) invited the Library Board to appoint a representative to its board, at that time not an elective office. Within a year Craver was picked by the ASME and the ASCE to represent them on the board.[72]

He had been active in the American Library Association for years, and had served on the ALA Council for two different terms, 1909-17 and 1924-31, four years of which included being on the ALA Executive Board (1913-17). He also had had various ALA committee assignments.[73] He served as president of ALA for the term 1937-38. Although the orientation of the Engineering Societies Library was technical, it was still a public library, as was Craver's previous library, the Carnegie Library of Pittsburgh. Thus his experience in two major public libraries was an appropriate library background for his high office in ALA. His presidential address was chiefly devoted to the need for better backup collections for small public libraries, many of which had few good sources from which to borrow books. (This turned out to be prophetic, inasmuch as the Engineering Societies Library became a backup source for the NYSILL plan some thirty years later.) He also spoke of hoping to see funding available for the hiring of readers' advisors.[74] One of the causes in which Craver became involved while head of ALA was his effort to persuade President Franklin D. Roosevelt to appoint a librarian as Librarian of Congress. However, Archibald MacLeish was eventually chosen for the position.[75] In 1939 Craver was made ALA's official representative at the meeting of the International Federation of Library Associations (IFLA) at the Hague.[76]

Craver also occupied a position of respect within academic circles, where his opinion about technical libraries was often sought. One example of this occurred in 1939 when Northwestern University needed advice about the size and type of collection for its proposed Technological Institute. He also served as a member of the Visiting Committee of the Massachusetts Institute of Technology, which needed his help to solve problems of overcrowded stacks and a library staff that was too small.[77]

Another facet of Craver's interests is seen in his paper on the library's role in engineering education, which he presented at a symposium held at Vanderbilt University in 1938. He touched on such factors as the need for better technical backgrounds for librarians in engineering libraries, the vital role of abstracting and indexing services, and the overall place of the library in scientific research.[78]

By the close of the 1930s Craver's career was at a peak by virtue of his successful year as ALA's president, his high standing among his colleagues, and his continued good relations with the Library Board at the Engineering Societies Library. However, in the coming years he was to be subjected to a hectic pace in directing the Library during the World War II period and to experience a decline in his health, leading to his ultimate retirement.

On 22 May 1942 a dinner was held in Craver's honor to mark his quarter of a century of service in the Library. At that time he was presented with an illuminated certificate of appreciation signed by the presidents of the four original Founder Societies. Craver used the occasion to reminisce about his entry into the library profession and his impressions of his library service. He pointed out that his library career dated back to a time when he had criticized the Carnegie Library of Pittsburgh for the inadequacies of its technical collection. As an account of his talk stated,

"As a result of this criticism he was asked to organize a technical library service, which he did, only to find himself a librarian."[79] His remarks were followed by a round of laudatory comments from Library Board members and officials of Founder Societies. In addition, letters praising his contribution to librarianship came from distinguished librarians and engineers.[80]

The collection of letters bound with the certificate from the board included one from Margaret Mann, who wrote about her appreciation for her fairness, his sense of humor, and his advocacy of high standards—qualities that she tried to pass along to her students at the University of Michigan. Ralph Shaw wrote in a similar vein. William Jacob, head of the main library of the General Electric Company, recalled how he, as an employee of the Engineering Societies Library under William Cutter, had been introduced to Craver by Cutter and how glowingly Cutter had described Craver, praise that was well justified. Charles Scott, a past president of the AIEE and one of the prime movers to obtain the Engineering Societies Building, recalled days in Pittsburgh (while Craver was still the technology librarian at the Carnegie Library) in which he and Craver had attempted to interest the AIEE in indexing technical papers in the institute's *Transactions* and distributing them in card form to libraries. Scott felt that the success of the *Engineering Index* in offering such a service was "substantial evidence that our ideas of forty years ago were sound." For the official record a resolution citing Craver's accomplishments during his twenty-five years of service appeared in the minutes of the Library Board's meeting of October 1942.[81]

In 1944, however, it became apparent that the time had come to find a replacement for Craver. His health was beginning to fail, and he was several years beyond the normal retirement age of sixty-five. The Executive Committee announced at its September meeting that it was appointing a search committee to find a suitable person to be trained as Craver's successor.[82] At a board meeting held in the spring of 1945 it was disclosed that Ralph H. Phelps had accepted the position as assistant to the director, to be paid $6,500, starting 1 July 1945.[83]

In January 1946 Phelps was named acting director and Craver was made consulting librarian. By that time Craver's health had deteriorated to the point that his work suffered badly. The plan provided for paying Craver one-half his previous annual salary. A farewell dinner was held in May 1946, and board members agreed upon a resolution citing Craver's accomplishments, read on the occasion of his retirement. A few of the achievements it listed were his establishment of a viable "Service Bureau" for paid service, the "Photostat Service," and a plan for making loans of books to members of Founder Societies. In the fall of the year Phelps was appointed director of the library.[84]

Craver continued to suffer poor health, and had little or no further activity at the Library. He died on 27 July 1951 at his home in Baltimore. One of the obituaries was written by Ralph Shaw, who recalled Craver's wit, incisiveness, and keen interest in Shaw's work.[85] Among the points Shaw made was that Craver was "patient with people while intolerant of second-rate work; a hard-headed visionary who constantly thought beyond tomorrow but simultaneously insisted upon practical planning of new projects; a fountainhead of knowledge on all phases of library work; and a philosopher of library administration who almost unconsciously required and stimulated everyone associated with him to grow."[86] On another occasion Shaw wrote that Louis Wilson, Harry Lydenberg, and Harrison W. Craver were his choices as the three top men of the library profession.[87]

Many of Craver's former employees, in the course of interviews, vouched for his decisiveness, his keen attention to duty, and his intellectual prowess. Yet some of them found him aloof and perhaps not very much concerned with the welfare of his employees. One felt Craver may have been ill at ease in dealing with employees and thus preferred to keep contacts at a minimum, while another person recalled his occasional mood of bantering over some incident. Still another librarian felt that Craver ran a one-man show and was a martinet. On the other hand, one former employee remembered Craver's compassionate efforts to get pension benefits adjusted to aid a staff member. Craver was generally respected as an able administrator, even if a feeling of warmth sometimes seemed to be lacking in his relationships with employees in his Library.

With the departure of Craver from the Library a major era came to an end—a period during which a loose confederation of four libraries had been welded into one cohesive unit by a librarian of ability, character, and national stature. At the end of Craver's term of office the Library had developed a collection of more than 175,000 volumes and had won status as an important technical library. During his administration of nearly thirty years, while enduring the complexities of leadership during two world wars and a devastating depression, Craver nurtured the reader service activities, reorganized the library staff, secured the services of a nationally known figure to oversee the recataloging of the collection, and fostered close working relationships with *Engineering Index*. It was a period of many accomplishments.

LIBRARY EMPLOYEES

Of the many employees of the Library during this period, two of them stood out as nationally and internationally prominent librarians. This chapter would not be complete without a brief summary of their contributions.

Margaret Mann

Margaret Mann was a natural choice for the role of supervising the recataloging project and creating a classified catalog. Like Craver, she was a native of the Midwest, born in 1873 in Cedar Rapids, Iowa. She was in the first class of the library school at the Armour Institute of Technology in Chicago. Following her graduation in 1896 she accepted the invitation to move to Urbana, Illinois, when the University of Illinois took over the Armour School, to be assistant librarian of the university library and instructor in library science.[88]

In 1903 she became the chief cataloger at the Carnegie Library of Pittsburgh, where one of her major accomplishments was the supervision of the final years of the preparation of a three-volume classified index to the holdings of that library as of 1902 (145,000 volumes). Published in 1907, it was the culmination of a project which had originated during the term of Edwin H. Anderson as librarian.[89] In 1908 a two-volume supplement was published, this being the set Craver sent to Andrew Carnegie soon after becoming librarian. Mann's work on the published catalogs attracted wide notice in the library profession. Years later W.W. Bishop wrote: "The Carnegie Library of Pittsburgh enjoys a peculiar distinction which it owes directly to Margaret Mann. It is the sole American Library of size and importance which has published a classified and annotated catalog on a large scale."[90]

Mann came to New York in 1919 and soon had the recataloging project underway while still keeping up with the cataloging of current books. She had a small staff, which makes even more noteworthy the amount of work they could accomplish, recataloging a total of 95,000 volumes in six years. The project was completed in 1925. Mann left before it was finished, as she had received a one-year leave of absence in 1924 in order to join the faculty of the Ecole des Bibliothécaires in Paris, a library school with American sponsorship which was part of the postwar effort by America to help rebuild war-ravaged French libraries. Her one-year leave turned into two years, and then she decided not to return to the Engineering Societies Library because she preferred a teaching career. She resigned from the Library in July 1926.

While Mann was at the Library she also taught part-time at the library school then operated by the New York Public Library.[91] This contact with the world of library education may have been a diversion that she needed as she worked on at the formidable job of the reclassification project. It could not have been an easy time for her. Years later, upon the occasion of Craver's completion of twenty-five years at the Library, she indicated how he had helped her: "You taught me how the highest standards of achievement take the drudgery out of effort and make work a pleasure.... Your sense of humor has carried me over many a problem that seemed insurmountable."[92] After her stay in Paris she was persuaded by William Warner Bishop to join his faculty at the newly formed library school at the University of Michigan, where she spent the final twelve years of her professional career. She died in 1960, with the reputation of being the best known instructor in technical services and of having written one of the most widely used American textbooks on cataloging.[93]

The Engineering Societies Library owes much of the excellence of its classified catalog to Margaret Mann's pioneering work in establishing the standards that would make the tool an accurate, valuable aid to library users for decades to come. Craver's choice of Mann to supervise the project was excellent, as those closely associated with the present catalog would readily confirm.

Ralph R. Shaw

Ralph Shaw worked at the library from 1929 (on a part-time basis at first) until 1936, at which point he was named senior assistant and chief bibliographer. Shaw was another Midwesterner, a native of Detroit. While earning his undergraduate degree at Western Reserve in Cleveland he worked part-time in the Department of Science and Technology of the Cleveland Public Library; he credited his later interest in science libraries to his experiences there.[94] He then received his bachelor's and master's degree at the School of Library Service at Columbia University, meanwhile working part-time at New York University. Later he switched to the Engineering Societies Library, where he did much of the research for his master's essay. It was a virtually unique study of early engineering works and became something of a bibliographic classic.[95] (His essay was subsequently reprinted in issues of the *Bulletin of the New York Public Library*.) Shaw's duties as chief bibliographer brought him into close contact with Craver, and the two got along well, as one library employee recalled.[96]

After he left the Engineering Societies Library to head the public library in Gary, Indiana, Shaw prepared a forty-page history of his former library, a work commissioned by the Library Board.[97] The section on the collection was particularly well done.

Shaw left Gary in 1939 to become head of the U.S. Department of Agriculture Library, where he remained until 1954. After that he became a library educator at Rutgers University and the University of Hawaii, where he died in 1972. During his career he proved his competence as inventor, bibliographer, library consultant, administrator, and publisher. His Rapid Selector—a device which "read" at high speed subject codes on microfilm copies of citations—was probably the invention that brought him his greatest attention, although it had a modest effect on the library profession.[98] Shaw was active in library associations, culminating in his election as president of the American Library

Association for the term 1956-57. He founded the successful Scarecrow Press. He was a dynamic and innovative man, although not universally liked. As one of his friends put it, "If some individuals found him egocentric and his wit sometimes too sharp and penetrating, many more found him to be fair, warm, generous, thoughtful, helpful and enormously loyal—to individuals and institutions alike."[99]

6

Notable Library Personages, 1946-80

This chapter describes the contributions of those responsible for the manner in which the Engineering Societies Library has developed in the post-World War II era up to the present.

Ralph H. Phelps (1946-68)

The directorship of Ralph Phelps covered twenty-two years, an important period for the Engineering Societies Library which included the adjustments to postwar conditions, the move to a new library building, and the involvement of the Library in regional and national projects aimed at improving the handling of information.

In some ways Ralph Phelps was a carbon copy of Harrision Craver. Both were products of the Midwest, both had undergraduate degrees in chemistry, and both had worked in the Technology Department of the Carnegie Library of Pittsburgh. But, the two men were entirely different in their personalities and their styles as administrators. Phelps felt that Craver's forte was "his ability to impress and influence his peers in other libraries, society secretaries and other officials."[1] Although the two worked together for about one year, Phelps felt that he never got to know Craver particularly well. He did become very much aware of Craver's skill with linguistics, a field in which he was "very gifted," in

Phelp's words. Comments from librarians who knew both men indicated a general feeling that Phelps was much warmer, much more approachable in his relations with his employees than Craver, who remained more aloof.[2]

Phelps was born on 5 July 1905, in Monmouth, Illinois, and went to Monmouth College in that town. He was graduated in 1928 with a degree in chemistry. Like many graduates, he had difficulty getting his first job because of a lack of experience, but that year he did find employment as a junior chemist at the U.S. Bureau of Mines in Pittsburgh. Before the year was up, however, he was working in the Technology Department of the Carnegie Library of Pittsburgh with the title of reference librarian. The head of the department, E. H. McClelland, had learned of Phelps and his background in chemistry through the Pittsburgh section of the American Chemical Society, and had offered him the library job. McClelland was more inclined to staff his department with persons having good technical backgrounds than with librarians, if he could not locate candidates having training in both subjects. After a few years, Phelps saw the advantages of having a library degree if he were to continue as a librarian, so he attended the Carnegie Library School part-time from 1934 to 1938. During his library school years he prepared abstracts for two technical periodicals, *Instruments* and *Metals and Alloys*. After receiving his library degree, Phelps went to work at the Birmingham (Alabama) Public Library, where he was in charge of the Science, Technology and Art Department. He had been aided in obtaining the position by his former employer, McClelland.

In 1942 Phelps became librarian for the War Metallurgy Committee, which was part of the National Academy of Sciences-National Research Council in Washington. He described the position as being head of a library with no books—the collection consisted almost entirely of security classified technical reports, and his chief duty was to review incoming documents and route them to appropriate staff members, a task dependent upon Phelp's evaluation of the subject matter of the reports. He recalled that the library was so cramped for space that a large Focault pendulum, which had formerly hung from the ceiling, had to be removed so that the space it used to swing across could be filled with shoulder-height wallboard panels used to create small offices or cubicles.

After three years or so at this job, Phelps was considered for the assistant director's position at the Engineering Societies Library. Craver had written to a number of technical librarians asking for recommendations of persons for the job. E. H. McClelland, Phelp's supervisor in Pittsburgh, strongly recommended him, as did his supervisor at the War Metallurgy Committee, Louis Jordan, who had formerly been the associate secretary of the AIME. Phelps was awarded the job and began work in New York in July 1945. He recalled that, with the exception of his first library job at Pittsburgh, he never had to apply for any job he accepted because people came to him with offers, although there were several jobs he did apply for and did not get.[3]

When Phelps began work in New York it was apparent that Craver was not well; he was out of the Library a good deal because of illness. When he was at work it was also apparent that he liked to spend a lot of time translating foreign technical articles and writing book reviews. Phelps felt that the smaller number of library users in those days and the smaller number of committee meetings helped explain why Craver had the time to work on such projects, unlike the experience Phelps had in later years.

On 1 February Phelps became acting director of the Library. His performance must have satisifed the Library Board, for within a few months he was named director, effective 17 October 1946.[4]

Administration, Staffing, and Governance   Before his first year at the Library was completed, Phelps had prepared a study of the past thirty years of growth of the Library and added an estimate for the next thirty years in regard to the need for space.[5] Another early task was to promote the Library as a resource for members to use. Several steps were taken; one was the approval of a new brochure for general use entitled "How to get engineering information." A letter was also authorized by the board to advise potential users of the cost of research services and to ask for donations for the Library.

In a related move, the board agreed to pay Phelp's expenses

in attending annual national meetings of the Founder Societies
in order to maintain better relations with them. Another sign of
the board's readiness to give Phelps more responsibility was its
appointment of him in June 1946 as a member of the United
Engineering Trustees (UET) Building Project Committee, an
early effort concerned with the prospects for a new building.
(The UET was the group to whom Phelps was responsible.) Still
another assignment came in November when he was sent to
persuade the McGraw-Hill Book Company to revive the Engineering Societies Monographs, a series that had paid royalties to the
Library before it lapsed. In April of 1947 the project was resumed,
and the Engineering Societies Monographs Committee was rein-
stated, with Phelps as chairman. By 1951 the Library had received
a total of $8,000 in royalties since the series began in 1931.[6,7] Like
Craver, Phelps received strong support from the board early in
his term of office.

A major change in the governance of the Library occurred
in October 1948 when the UET announced a new organization
plan for the Library Board. Rather than having a board consisting
of twenty-one members, the new plan called for a reduction in the
size to fourteen members (thirteen to have a vote), consisting of
two representatives from each of the four Founder Societies, two
representatives of the UET board, two representatives of all
Associate Societies (those not legally bound to be located in the
UET building), the president of the UET Board of Trustees, and
the director of the Library as the nonvoting secretary. There was
also provision for two alternate members, representing the
Associate Societies, who could act and vote only in the absence of
the regular representatives of these groups. A month later it was
decided to abolish the Library Board's Executive Committee and
to have the full Library Board meet eight times per year rather
than only four. This latter move and the new organizational plan
simplified the operation of the board, while still keeping it representative of various constituencies. A new set of rules of administration for the Library was also adopted at this time, subject
to UET approval; this document specified the names and duties of
committees, the number of meetings of the board, and the like.
The UET approval was subsequently granted.[8]

An important document was revised in August 1950 when it was decided that the time had come to update the Library Agreement dated 1 January 1915, as amended in 1916 when the ASCE became a Founder Society. The first clause of the new agreement made it clear that the UET claimed ownership of all the literature, stacks, furniture "and all other movable things in the custody" of the Library; thus the disagreements over this point which had plagued the early years of the Library were eliminated by the clarity of the wording of this item in the document.

In 1958 the American Institute of Chemical Engineering (AIChE) was accepted as the fifth Founder Society, effective 1 May 1958, an addition of thousands of members to support the Library. While there was practically no public fanfare over this event as compared to the addition of the ASCE, there was genuine gratification that another major society had joined the UET.[9]

During Phelp's term the use of the Library continued to climb, with telephone inquiries showing huge increases in certain years. For example, annual reports show that 1946/47 there were 7,400 such inquiries, whereas in the fiscal year 1959/60 the number had soared to almost 25,000, remaining generally above the 20,000 mark thereafter. Likewise the photocopy service showed large annual gains; for instance, in 1946/47 a total of 48,000 exposures were made as contrasted with 87,000 in 1959/60 and a mind-boggling 238,000 in 1967/68. Phelps tried to keep charges as low as possible consistent with meeting costs of providing the service. Concerned about employees, he persuaded the Library Board to grant raises in March 1953, by pointing out that salaries averaged $3,441 in contrast to $4,092 at other libraries, or $3,528 for starting handymen at the Engineering Societies Building.

One of the oustanding characteristics of Phelp's term of office was his interest in cooperating with other libraries, coupled with an ability to deal with all types of people in a rational, friendly fashion. For example, early in his term he willingly met with officials of the New York Public Library (at their invitation) in the first of several attempts to see if more cooperation could be reached in acquisition and collection practices, a topic he took up a few months later with representatives from local technical libraries. While no particular progress was made, it indicated his interest in working on agreements of this sort with others.[10,11] In subsequent years he spent countless hours attending meetings sponsored by either librarians or engineers to discuss various projects involving the Library. No doubt his outlook on attempts to reach agreements led to his heavy involvement in this sort of activity; a person with no interest in or patience for collective agreement would soon have dropped out in view of the glacially slow progress such committees and groups usually experienced.

Committees and Projects   Phelps did not become officially involved with *Engineering Index* until he had been director of the Library for nearly six years; then, in 1955, he was named to its Board of Trustees, a group of up to fifty people who met annually and elected the Board of Directors from among themselves. Phelps was also elected to the Board of Directors of the *Index* in 1955, and continued as a member until 1968. He kept his membership on the Board of Trustees until 1974. Further recognition came to him in 1956 when he began a six-year term as vice-president of the Board of Directors, then followed with six years as its president (1962-68). So it is evident he had a large role to play in the affairs of the *Index* for many years.[12]

During his years on the *Index's* Board of Directors he saw to it that full cooperation was maintained between the Library and the *Index*, such as provision of access to the Library's literature for indexing purposes and consultation with *Index* staff members in the selection of new literature for the Library's collection. Practically all of the Library's journals and many of its monographs continued to be covered by *Engineering Index*.

One of the many committees on which Phelps served was the Church Committee, organized in 1946 to study on a long-range basis the operation of the Library and its objectives. Its report noted that the Library had become overcrowded, it was operating at a deficit (using a surplus gained in 1948/49 to balance its budget), and it needed to improve salaries. Recommendations were made to correct these problems. In addition to attention to matters of this sort the report, issued in 1947, also looked into the future.

> The Committee envisages a future well-publicized "Master Engineering Library" which shall serve not only a fraction of the 80,000 Founder Society members

but a consideration fraction of the 260,000 engineers in the United States. Such a master library would contain a far more complete collection of engineering books and periodicals, American and foreign, and would provide adequate library research and mail service to all parts of the country.[13]

There was no immediate reaction to this report.

In the 1960s Phelps became heavily involved in analyzing the proper role of the Library in automation, a topic of keen interest to the engineers on his board as well as to him and to many librarians. The general tenor of the 1960s was that of re-examining traditional ways of doing things, especially in the light of the possibilities offered by computers. The reevaluation process was healthy, but there were unfortunate projects implemented in other institutions that became costly failures. Phelps and his colleagues at the Library were aware of the pitfalls of unwise automation applications, yet they had to remain open to the possible benefits of new methods and systems. It was not an easy time for Phelps. He recalled later that many times during those years he left the Library at night deeply disturbed about misuse of automation. Part of his problem stemmed from the fact that some of the board members and others working with him had more enthusiasm than wisdom about automation and systems analysis.[14]

At that time one of the groups urging the Library Board to get more involved in automation was the Engineers Joint Council, a unit consisting of representatives of nearly all the engineering societies in the United States. Another spur to action came from Raymond P. Genereaux, a board member who in 1962 urged the Library to consider a large indexing project leading to computerized searches.[15] By the spring of 1963 a rather confusing array of committees and task forces had been created to study automation in relation to the Library and to *Engineering Index*. In that year another committee, known as the Coordinating Committee, was established for the purpose of reviewing the various proposals and making recommendations for groups such as the Engineering Foundation, which had also become involved. Phelps served on many of the committees, including this one. Then in 1965 the

Bowie Committee urged the creation of a nationwide engineering information network involving the Library and *Engineering Index*.[16] This was partially superseded by the creation later that year of what became known as the Tripartite Committee, made up of representatives of *Engineering Index*, the Engineers Joint Council, and the United Engineering Trustees.[17] It was to include high officials of these groups in an effort to discourage the pro-; liferation of smaller, less prominent committees and study groups that had been so evident in those years. Armed with funds from the National Science Foundation, the Tripartite Committee, after much deliberation, asked the Battelle Memorial Institute to make a study leading to a plan for a nationwide information system; work began there in 1968 and was still underway when Phelps retired later that year.[18]

The New Building   One of the major accomplishments of Ralph Phelps was his active role in the planning of and the move into a new library in the early 1960s.

After the American Society of Civil Engineers (ASCE) moved into the enlarged Engineering Societies Building in 1917, the headquarters of the four largest engineering societies in the United States were under one roof, with a joint library to serve them and their members. Ideally, this should have been a long-term solution to their need for working space, but the building began to be too small for its occupants. One reason was the rapid increase in membership of the societies, with a corresponding need for a larger headquarters staff. For example, the ASCE in 1915 had 7,707 members; in 1935 it had 14,218, nearly doubling its size in twenty years. By 1955 its membership (38,169) had more than doubled again.[19] Membership in the other Founder Societies can be estimated from the great increases in the number of engineers, whether or not affiliated with societies, as seen in Table 1; for example, in 1910 there were only 77,000 engineers of all types, but by 1950 there were over half a million.

Another problem was caused by the fact that engineering literature continued to grow apace during those decades, so that the collection gradually outgrew the stack area. A much-needed addition to the stacks on the fourteenth floor was finally completed, long overdue, in 1949; this $7,000 project gave the Library space for an additional 10,000 volumes, barely enough for four years of growth. The addition had been sought in 1937, at which time the stacks had already become too crowded for efficient use, so one can imagine how bad the shelving situation had become by the completion date of the project, twelve years later.[20]

As early as 1928 consideration was being given in some quarters to the need for a new building. In 1942 a practical step towards a future move was taken by the UET when it sought and obtained an amendment to its state charter allowing the sale of its Thirty-ninth Street building as long as the proceeds were used for the acquisition or erection of a new building to be utilized for the same purposes.[21]

There were brief references to the possibilities of a new building here and there in official records, but it was not until 1954 that definite action was taken. At the February board meeting a report from the UET on the proposed new building was read. In order to acquaint the Library Board with the existing space occupied by the Library, Phelps gave the board a tour of the Library and reviewed the space needs of the Library back to 1907. The Board asked for a total of 26,600 square feet, which was estimated to be enough space for growth until 1980.[22]

After considering possible sites in other cities, the decision was made to build in New York City, the center of heavy concentration of engineers in the Northeast and the headquarters of important engineering firms. Another consideration was the wording of the deed turning the Wheeler gift over to the Library; if the UET offices and the Library had left the city, this collection would have had to remain in New York. Since the volumes were interspersed throughout the collection, it would have been a time-consuming task to locate them. And former President Hoover personally urged that the building be built in New York City.[23]

The UET Board declared that tenants should be restricted to "engineering societies or related organizations having the same classification as educational or scientific organizations which the Founder Societies and UET now enjoy." This meant that the proximity of the Engineers' Club to the society headquarters building would end unless the club found quarters of its own near the new center (which it never did).[24]

Funding for the new building took a great step forward with the signing of an agreement between the UET and the five Founder Societies whereby the latter would each loan from $250,000 to more than $325,000 to the UET for the purchase of a site on First Avenue, between Forty-seventh and Forty-eighth Streets, on which would be erected a building to be known as the United Engineering Center. This transaction was to be independent of any contributions they might wish to give to the building fund later. Another agreement reaffirmed ownership of library materials by UET and set forth the obligation of the societies to support the Library financially.[25]

In June 1958 the Library Board sent a request to UET for the following space in the new center, this being the amount it estimated would be needed up to 1980:

Staff areas	4,600 sq. ft.
Reading room area	4,500
Stacks	18,000
Total	27,000 sq. ft.

The library ultimately received this amount plus a few hundred square feet more, a great improvement over the 10,000 square feet which UET had first proposed for the Library.[26]

At the time of purchase the site was quite drab, containing a garage and a service station, not far from some grimy three-story buildings. It was in what is still called the Turtle Bay area.[27] Even though the attractive United Nations buildings, in existence since the early 1950s, had brought a new elegance to what had once been a ugly industrial district, and though there was a building program underway in the immediate vicinity of the UN area, there were still several unattractive sections of the neighborhood that were untouched by the upgrading process. However, the UET decided there was little risk that any property in the vicinity of the UN buildings would not increase in value, and the selection of the site chosen would appear to have been a good real estate investment. Today when one views the United Nations structures, the modern apartment buildings across the street, the handsome United Engineering Center, and the many other new structures in the vicinity, it is difficult to imagine how unattractive the site used to be.

In September 1958 the Library Board appointed the Library Committee on the New Building to help draw up long-range plans for moving the Library and for computing the costs of the new facility, among other topics. Ralph Phelps recalled later that since he feels he is a designer at heart the process of doing the planning for the new library was "more fun than I've ever had in my life."[28]

Phelps related a little-known incident in the purchase of the land for the new building. It seemed that after the various lots had been purchased by a dummy corporation, the device often used to allow the acquisition of a number of parcels of land without prices being unduly increased, it was discovered that there was a four-inch strip of land which no one owned. It took joint action by the UET lawyers and various city officials, including Mayor Robert F. Wagner, to solve the problem and add the strip to the UET holdings.[29]

Construction moved along quickly, once the contract for the building had been signed in May 1959. The ground-breaking ceremony was held on 1 October of that year, with Herbert Hoover, a former mining engineer himself, participating. He said that engineers "are the foundation of security in our defense," and pointed out their role in "the increase of our standards of living and comfort."[30] On 16 June 1960 the cornerstone setting took place, with Hoover making a speech entitled "A Message for Posterity," a copy of which was placed in the cornerstone along with publications of societies. Speaking on the evils of communism, Hoover reminded the 450 or so attendees at the ceremony that progress in the last century came from "free and productive minds in the civilized parts of the world—that is, the nations possessed of independence and personal freedom."[31] He said, "This building is to facilitate the goals of the engineer.... It will play a great part in American life. It will serve all mankind.[32] His talk was preceded by brief remarks from Mayor Wagner, who praised the decision to build the center in New York City. He related how city officials had helped find the ultimate site for the building and gave other assistance, all stemming from a meeting held in April 1955 when the societies were considering the question of moving. Wagner also made public the fact that Hoover had sent the UET a letter urging them to remain in New York City because of the intellectual climate it offered. At one point Wagner said, "Our tribute is not to a building because it will enhance the skyline of our City but because it will accommodate a class of professionally trained men and women who are dedicated with a common, unselfish interest; the benefit of their fellow men."[33]

By September 1961 the twenty-story building was completed with the exceptionof interior work on two floors, for which there were no tenants at that time. (These two floors were occupied in 1964 by the Institute of Electrical and Electronics Engineers [IEEE], an organization formed in 1963 by the merger of the American Institute of Electrical Engineers [AIEE] and the Institute of Radio Engineers [IRE]. This merger brought an additional 88,000 engineers, and thus more financial support, to the UET.[34]) The finished office space was occupied almost as soon as the building was completed. The formal dedication, held on 9 November 1961, was a gala occasion, complete with an Air Force band. Greetings were received from President John F. Kennedy and Governor Nelson Rockefeller. Mayor Wagner noted that the center represented the largest cooperative gathering of professional engineering organizations in the world.[35] Herbert Hoover was again an obvious choice to deliver remarks at the completion of a building which would house nineteen professional engineering societies and related engineering agencies, representing over three hundred thousand engineers. He pointed out that the United States had fallen behind the Communists in the number of scientists and engineers produced, but said the remedy could be provided by a "united front" of engineers, a unity that was embodied in the new $12.5 million center.[36]

There was general pleasure among the Library's staff and users over the new building, which was twice the size of the Engineering Societies Building and included space for meeting rooms and headquarters for 50 percent more societies than could be fitted in the former building. The cost of the building had been partially covered by a very successful fund drive; for example, an Industry Committee under the chairmanship of Dr. Mervin Kelly raised $5 million through contributions from industry. Many thousands of individual engineers made personal contributions, and the societies devoted space in their periodicals to soliciting and reporting on such gifts. Altogether, total contributions amounted to over $5.2 million from industry, over $4 million from members and $300,000 from the Ford Foundation. Added to this total was income from the sale of the old Engineering Societies Building in October 1960 for $2 million, about one-third of its replacement cost. (The buyer's hope to see it used for the garment industry or for professional/educational purposes never materialized, and it became a public garage in the lower floors and miscellaneous offices above.)[37]

Phelps found the new Library a great improvement over the old. The stacks, for example, were easily accessible from the reference desk area as contrasted to the dark, almost inaccessible stack floors in the old building. The new reading room was in clear view from the reference desk. There were well-planned staff areas, with room for the photocopying and microfilming operations. This satisfactory layout was due to much planning on the part of Phelps, who had also frequently requested suggestions from the staff. The 27,465 square feet the Library had been granted gave it attractive staff and user areas plus ample growth room for the collection. It was a different world from the crowded, poorly planned library they were leaving behind.[38]

The general satisfaction did not mean there were no problems with the new library. For one thing, the location was not so convenient to public transportation for many staff members and users. There was no provision for a public coat room, so that one alcove in the reading room had to be used for that purpose and for housing the self-operated photocopy machines. The new reading room's tile floor proved to be too noisy and eventually had to be carpeted.[39] These were, however, all relatively minor problems, of far less import than the good features of the new library.

The library staff had made many preparations for the move, which took place from 12 August to 5 September 1961, work con-

The United Engineering Center.

tinuing on a fourteen-hour-per-day basis during that period.[40] Although the reading room was closed to readers during the move, the staff was still able to answer mail requests for loans and photocopies. Some 180,000 volumes and 700,000 catalog cards were transferred. Considerable weeding had been done in advance: for example, all of the 18,000 topographic maps from the U.S. Geological Survey were disposed of prior to the move, since the New York Public Library already had a strong collection. There was a professional staff member on hand at all times to supervise the move. In general the transfer of books was handled well, with little refiling necessary afterwards.[41] The move cost about $24,000, in contrast to the original estimate of $14,000.[42]

The library staff had grown slightly in size, from twenty-six persons in 1947 to a total of thirty in 1961, half of whom were professional librarians. Phelps found that his time was being taken up more and more with committee meetings associated with the Library, making it harder for him to give proper atten-

The Library reading room, United Engineering Center.

tion to the routine duties of daily operations. By 1963 the matter had become so serious that the Board considered creating a new position in the Library, that of assistant to the director. At the fall meeting when the budget for the 1963/64 fiscal year was being prepared, the board decided to include money for the proposed post, and Phelps began to interview candiates.[43] At the January 1964 meeting it was announced that S.K. (Kirk) Cabeen had been hired, to start 1 February 1964; as of 20 February he began attending board meetings with Phelps.[44]

Phelps faced many administrative problems, such as seeking ways to increase library funds. *Engineering Index* was also undergoing financial problems in the mid-1960s, and the possibility of merging the Library and the *Index* was once again considered and rejected.[45] Still other problems involved securing raises for employees to match the equivalent salaries paid employees at the New York Public Library, as well as instituting better fringe benefits for employees.[46]

Phelps's interest in expanding the usefulness of the Library led him in the winter of 1967 to enroll it as a member of METRO, an organization of libraries of all types in the New York metropolitan area. At that time he announced the Library was considering an offer of a contract with the New York State Library to participate in an interlibrary loan plan whereby the Library would act as one of seven subject backup libraries. This move, formalized in 1969, was to aid New York libraries in a very positive fashion and also to provide the Engineering Societies Library with welcome additional funds.[47]

Another administrative decision involved appointing Marguerite Soroka, head of the Cataloging Department, as head of the Technical Services Department, merging cataloging and acquistions activities into one unit.[48] Table 2 shows the organization chart for the Library as of 1966.

Under Phelps the number of serials being received ranged from some 1,400 in 1954/55 to 3,400 in 1978, with considerable success in obtaining free subscriptions of major journals.[49] Before he left the library Phelps saw the collection total more than 200,000.[50]

TABLE 2
LIBRARY ORGANIZATION CHART, 1966

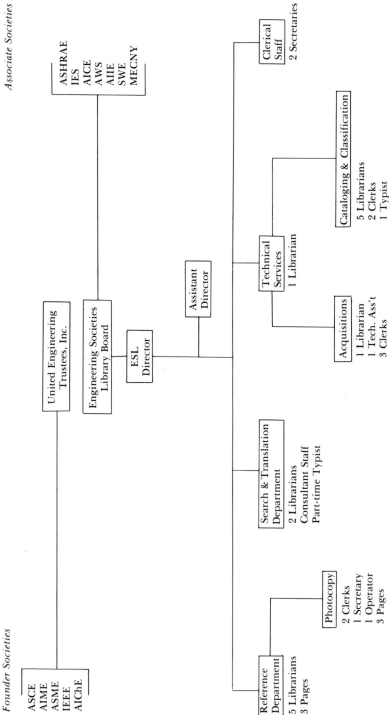

Founder Societies

ASCE
AIME
ASME
IEEE
AIChE

Associate Societies

ASHRAE
IES
AICE
AWS
AIIE
SWE
MECNY

United Engineering Trustees, Inc.

Engineering Societies Library Board

ESL Director

Assistant Director

Reference Department
5 Librarians
3 Pages

Photocopy
2 Clerks
1 Secretary
1 Operator
3 Pages

Search & Translation Department
2 Librarians
Consultant Staff
Part-time Typist

Technical Services
1 Librarian

Acquisitions
1 Librarian
1 Tech. Ass't
3 Clerks

Cataloging & Classification
5 Librarians
2 Clerks
1 Typist

Clerical Staff
2 Secretaries

Phelps's Professional Development   During his career Phelps had been active professionally, holding several offices in various library organizations and engineering groups. The Special Libraries Association was the library organization in which he made his largest contribution, serving at the local chapter and association level. He was treasurer of SLA (1962-64) and was a member and/or chairman of a variety of SLA committees.

Besides participating in library associations Phelps was heavily involved in *Engineering Index* affairs, as described in the previous chapter, including being on the Tripartite Committee from 1965 until his retirement in the fall of 1968. As one of his colleagues and fellow Tripartite Committee members, Eugene B. Jackson, recalled, Phelps maintained a willingness to listen to proposals to the Tripartite Committee for making improvements in information handling while being quite aware of financial implications. He tried to serve as a mediator between those committee members who thought the tripartite decisions were too radical, and those who considered them too conservative. Because of Phelps's efforts on the Tripartite Committee as well as his many years of service for more than a decade as vice-president and president of *Engineering Index*, Jackson is of the opinion that Phelps's contributions to the *Index* were even greater than his accomplishments in the Library, though he was much better known for this work with the latter.[51]

Another organization to which Phelps gave of his time and effort was the American Standards Association, whose Z39 Committee was involved in the preparation of standards pertaining to library matters. For example, he served as its chairman in 1954-55, as well as heading a subcommittee concerned with the layout of periodicals in 1951-52 and 1955-56.[52]

Phelps was also a rather prolific author of papers for professional library and engineering journals. One article, entitled "Engineering information—all is not lost," was so well received that it appeared in at least two periodicals and one collection of articles after its original presentation at an annual ASME meeting.[53] The theme of the article was that the Engineering Societies Library, in conjunction with *Engineering Index*, was serving thousands of engineers, and at a reasonable cost. As a consequence of his article, Phelps received letters of praise from many colleagues, including E.J. Crane, director and editor of *Chemical Abstracts*, and Verner W. Clapp, president of the Council on Library Resources. Crane's opinion was that Phelps's article was needed because it pointed out that the workable methods of traditional information organizations should not be tossed aside for untried methods until new ones were found worth accepting: "No method is old fashioned until a better method is found."[54] Phelps's other writings covered a broad range of topics, including cataloging, reader services, and selection policies. Many of his articles, written in order to encourage the use of the Library, appeared in journals of sponsoring societies.

Phelps' Retirement At the January 1968 Library Board meeting it was announced that Phelps had decided to take early retirement effective the following September, when he would be past his sixty-third birthday. A search committee was named at the February meeting and began its deliberations. Cabeen, who had been assistant director since December 1965, was an obvious candidate for the opening, and was appointed Phelps's successor effective 1 October 1968. The nomination was ratified by the UET Board at its April meeting. Efforts were made during the remainder of Phelps's term to hire a new assistant to the director, but one promising candidate decided at the last moment to take another position. Other candidates were not generally qualified, so Phelps's retirement came without a replacement selected for Cabeen.[55]

All of the people interviewed who had been employees or colleagues of Ralph Phelps had nothing but praise for him and for his genuine regard for all those with whom he worked. Employees found him fair and concerned about their welfare; his peers regarded him as a diplomatic person who nevertheless would speak out for what he believed in. As one colleague said, Ralph Phelps never showed the least bit of anger at any time, no matter how vexing the situation might have been; he knew how to present his point of view without upsetting people.[56]

S.K. (Kirk) Cabeen (1968-   )

S.K. Cabeen was, unlike his two Midwestern predecessors, of Northeastern origin. He was born in Easton, Pennsylvania, on 22 January 1931. Like Phelps, Cabeen attended college in his home town, in his case Lafayette College. Working in the college library provided the spark that made him decide to become a librarian. He received a bachelor's degree in chemistry in 1952, thus continuing the unbroken string of directors from Cutter through Phelps who had a degree in chemistry. Next he went to Syracuse University's library school, which granted him the master's degree in 1954. After a two-year stint in the Army Chemical Corps, Cabeen began his library career in 1956 as assistant librarian at American Metal Company, Ltd. (now AMAX), in New York City. He remained there until 1958, then worked for six years at the Ford Instrument Company, a division of Sperry Rand Corporation, at Long Island City, New York, where his title was technical librarian.[57]

Cabeen's service at the Library dated from his being hired as assistant to the director as of 1 February 1964, followed by his promotion to assistant director in December 1965. He was a natural choice to be Phelps's successor when the latter retired at the end of September 1968; announcement of this promotion was made about four months ahead of Phelps's departure.[58] Cabeen had been active in the Special Libraries Association; he had been president of the New York chapter for the year 1966/67 and served in numerous other positions at the chapter, division, and association level. After his appointment, he continued to be active in SLA as well as in the American Society for Information Science.[59] It should be noted that the longevity of his predecessors was such that he became only the fourth director of the Library since its formal creation in 1913.

Administration and Finance   There have been relatively few significant staffing problems in Cabeen's term; the long tenure of many of the staff members lessened the frequency of filling vacancies. The main position he had to fill was that of deputy director, a position he created in order to have a person in charge of the Library in his absences. His choice of Marguerite Soroka, since 1965 head of the Technical Services Department, was approved by the board and announced at its January 1972 meeting.[60] Soroka had been employed at the Library since 1946, rising from cataloger to head of the Cataloging Department in 1958.[61] Her new duties included consulting with the director in assigned

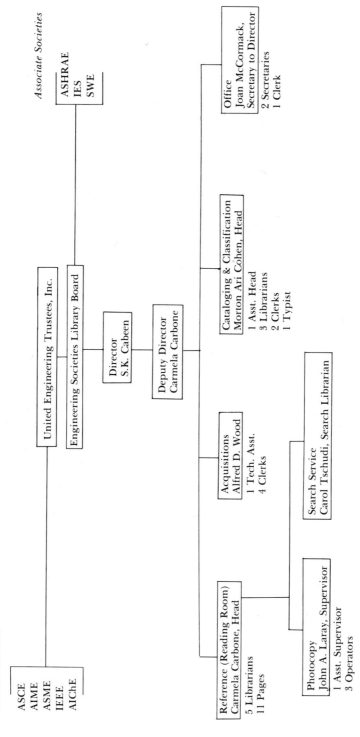

TABLE 3
LIBRARY ORGANIZATION CHART, 1980

Founder Societies

ASCE
AIME
ASME
IEEE
AIChE

Associate Societies

ASHRAE
IES
SWE

United Engineering Trustees, Inc.

Engineering Societies Library Board

Director
S.K. Cabeen

Deputy Director
Carmela Carbone

Cataloging & Classification
Morton Ari Cohen, Head
1 Asst. Head
3 Librarians
2 Clerks
1 Typist

Office
Joan McCormack,
Secretary to Director
2 Secretaries
1 Clerk

Acquisitions
Alfred D. Wood
1 Tech. Asst.
4 Clerks

Search Service
Carol Tschudi, Search Librarian

Reference (Reading Room)
Carmela Carbone, Head
5 Librarians
11 Pages

Photocopy
John A. Laray, Supervisor
1 Asst. Supervisor
3 Operators

aspects of the Library's administration as well as serving as director in Cabeen's absence. She retained her duties as head of the Technical Services Department.

However, Cabeen had the same position to fill when Soroka retired in April 1980, bringing to a close an exemplary career. Cabeen selected as the new deputy director Carmela Carbone, who also retained the title of head of the Reference Department, a responsibility she was assigned in 1978.[62] In general, the staff has remained remarkably unchanged during Cabeen's term as director. Table 3 shows the organization chart of the Library in effect in 1980.

One problem Cabeen did have in common with his predecessors was that of finances, since the Library had traditionally faced budgetary restrictions. The advantages of occasionally seeking an increase in allotment paid annually for each member of sponsoring societies had been recognized by each of the Library's directors as an important source of extra income. For example, by 1970 the total number of members involved was in the neighborhood of three hundred thousand, so that even a small increase would make a welcome addition to the budget. In the early 1970s Cabeen and the Library Board were able to persuade the UET Board to increase the allotment a modest ten cents per member, bringing the annual figure to sixty cents per member.[63] However, at the same time an extra $30,000 UET had been paying annually was discontinued, so there was no net gain to speak of. Cabeen was aware that the income from members, long the mainstay of the budget, was gradually being displaced from its primary rank by other sources. In 1972 he showed the board a chart that made this clear (see Table 4).

He redoubled his efforts in the seventies to find other ways of increasing Library income. For example, considerable effort was made to seek voluntary contributions to the Library from corporations. Company memberships, which were instituted in 1964, allowed for the borrowing of books by company libraries and brought in additional income, even if the charge was at the relatively modest rate of $100 per year or per 100 books borrowed. By the fall of 1980 there were more than sixty such memberships.

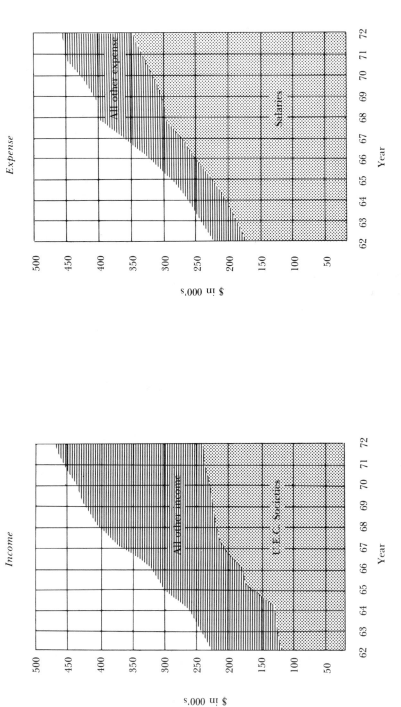

TABLE 4
LIBRARY INCOME AND EXPENSES, 1962-72.

Another new source of income appeared in 1970, coming from a coin-operated photocopy machine installed in the reading room for the use of the public. By 1980 there were three such machines, providing an annual income of more than $40,000. By the fall of 1980 the income from photocopies made by the Library had increased to approximately $561,000 per year. The income from the NYSILL contract to provide requested materials on a fee basis continued; by the fall of 1980 the total cumulative amount since the program began in 1967 was more than a quarter of a million dollars.[64],[65] The assessment for the more than 400,000 members of sponsoring societies is now sixty-five cents per member per year; this five-cent increase currently adds approximately $20,000 per year to the Library's funds.

Cabeen's efforts to increase nonsociety income were bearing fruit, but it was still a tight budget within which to operate. Expenses had to be very carefully watched; for example, expenditures for salaries, which constituted the largest single item in the expense budget, were made as low as possible by keeping a tight rein on the size of the staff. In 1962, to use one comparison, there were twenty-eight permanent employees in a year in which 8,000 items were added to the collection; in 1972 there were only twenty-nine employees for a year in which 13,400 items were added. Thus it is apparent the Library operated with a lean staff, as far as total number was concerned.[66]

Another vexing financial problem arose in 1971 when New York City proposed to levy a real estate tax on the United Engineering Center, which up to then had been considered tax-exempt. UET officials persuaded the city to drop its plan to tax the property. The problem arose again in 1975, and once again the plan was dropped. In the interim between the two tax threats the UET Board made some studies about the advisability of moving from New York City, but the decision was made to remain in the city.

Lest it be thought that financial woes were confined to the Engineering Societies Library, there were similar problems at many other technical libraries. The New York Public Library's Science and Technology Division gave notice late in the fall of 1971 that it was in financial trouble, and narrowly avoided being permanently closed to the public.[67]

Committees and Projects   When Cabeen's term as director begin, an important report was almost completed for the Tripartite Committee, the study group which had been established in 1965 to investigate the possibilities of creating a national engineering information system. It had representatives from *Engineering Index*, Engineers Joint Council, and the UET itself. The committee had commissioned the Battelle Memorial Institute, in Columbus, Ohio, to prepare a detailed plan; an informal summary of the long-awaited report from Battelle to the Tripartite Committee was finally received in May 1969 and discussed at the meeting of the Library Board that month. The plan proposed the creation of a United Engineering Information Service which would be involved in such activities as planning and sponsoring research in information science, the marketing of technical information, the operation of a referral service (to refer questioners to appropriate agencies or libraries), the publication of a newsletter about technical information, and the preparation of state-of-the-art reviews of technical subjects. There was no provision for a role for the Library or for *Engineering Index*, for the creation of an indexing and abstracting service, or for the operation of a document repository. In general, the response of the Library Board was not enthusiastic, particularly in view of the long period of time and large amount of money used for the study. A formal abstract of the Battelle plan was discussed at the board's October 1969 meeting, at which time a resolution was adopted that the plan was of dubious value and that the Library would have continued interest "in the distribution of engineering information."[68]

The plan was presented to the public at an invitational meeting in the United Engineering Center and drew more criticism than praise. Only the general nature of the Battelle plan was described in its "Action Plan," which had been released towards the end of 1969.[69] It listed three main areas in which to take action: being a unifying force in the engineering community by, for instance, coordinating the work of various engineering societies with that of the government; conducting studies on the use of engineering information and offering educational courses; and keeping engineering managers informed of developments in the information field by means of newsletters and critical reviews of the literature. A more detailed account of the work of the Battelle project and its recommendations were given in a paper by Liston, one of the project managers.[70]

While there was no particular objection to the usefulness of the various recommendations described in the report, the general reaction seemed to be that the study was not very significant—its suggested plan of action fell far short of what was expected of it. A former member of the Tripartite Committee, Ben Weil, felt quite disappointed over the scope of the Battelle Report. He had served on the committee for its first two years alongside Ralph Phelps, as vice-president and president, respectively, of *Engineering Index*. These two found that the governing boards of both the *Engineering Index* and the Engineering Societies Library favored a plan whereby the two parent organizations would be merged. This concept had been discussed widely, but the Tripartite Committee disregarded the idea in favor of asking for the Battelle report. (Batelle had won the contract in a competitive process.) As Weil described the Tripartite Committee, "The mountain labored and brought forth a mouse, and the community stepped on the mouse."[71] As there was almost no favorable response after the disclosure of the final report, no definite steps were ever taken to implement the proposals Battelle set forth.

Despite the lack of progress in defining new roles for the Library in relation to *Engineering Index*, close relationships continued between the two organizations. After a computerized version of the *Index* was created in 1969, the Library offered on-line search service for the new database and others then becoming available. This service was only fully realized after the reorganization of the Reference Department in 1980.

After the Tripartite Committee report was put to rest, a variety of studies and reports were created over the next few years, mostly proposing different rules for the Library and/or *Engineering Index*. Nothing concrete came of any of them, but one study stood out as having a broader scope than the others: the North Study, emanating from a proposal made by Library Board Chairman R.A. North in 1971, aimed at developing a fresh look at what services and activities the Library should be engaged in, as viewed by Library Board members.[72]

Despite the failure of the various study groups to evolve an innovative sort of connection between the Library and *Engineering Index*, relationships between the two organizations continued good, with wholehearted cooperation obviously in evidence in all phases of their activities. In the closing months of 1980 *Engineering Index* made ready for a public announcement of its intention to offer literature searches on a fee basis , creating an apparent competitive service to that of the Library. Only time will tell the outcome of this development.

Cabeen held the same open mind towards cooperation with other libraries that Phelps had shown. The most significant project was the highly successful NYSILL plan, under which the Library became one of the twelve libraries in New York State to serve as a backup for the interlibrary loan system operated by the New York State Education Department. Under the plan, each library among the twelve is paid a fee by the state for each request handled from a library in New York plus additional money for each request successfully filled. This contract put the Library directly in a major role in an important, useful network, and by the fall of 1980 the cumulative total it had earned since 1967 amounted to more than $365,000.[73]

Services to Library Users   As previously mentioned, the one service which experienced the greatest growth was that of photocopying items in the collection, ranging from a low point of 200,000 pages copied for requestors in 1972/73 to a high point of more than 613,000 in 1979/80. In the reading room, pages copied in the period increased to 165,000 by 1979/80.[74]

All this was done with careful attention to the revised federal restrictions on copyright. As for the number of requests for service, the range was from a low of 67,000 in 1972/73 to more than 128,000 in 1979/80; nearly 100,000 requests in the latter year were from telephone or mail contacts. This illustrates the growing use of the Library by people outside the New York City area. In fact, a study made in 1978 showed that only 3.2 percent of the photocopy orders originated from New York City.[75] There is every indication this geographically broad-based use of the Library is continuing.

Another fast-growing service is one that developed in the fall of 1980, when the Library became an officially recognized source of copies of literature indexed on databases offered customers of the Lockheed Information Systems, whereby requests are sent by customers from their terminal via Lockheed to the Library. By the end of 1980 there were an average of twenty requests per day. In addition the number of searches that were done by the Library using its terminal to contact the Lockheed databases has been gradually increasing since the spring of 1980. The cost of searches has ranged from no more than twenty-five to several hundred dollars, depending upon the complexity of the search and the amount of material available to be searched. Both these developments indicate that the Library is keeping abreast of the times in view of the immense popularity of on-line searching of computerized databases.[76]

Growth of the Collection   When Cabeen became director in 1968, the collection totalled approximately 217,000 volumes and 10,000 maps. By the fall of 1980 it had increased to more than 255,000 volumes and 12,000 maps. In 1968 the Library was receiving approximately 4,400 serials; by 1980 it had reached the mark of more than 7,400 titles, coming from fifty countries in twenty-five languages.[77]

LIBRARY BOARD

Of all the many men (and a handful of women) who served on the Library Board over the years, one person made an outstanding contribution and thus should properly be included in this chapter. Regretfully many others who served long years and gave unstinting support to the Library cannot be mentioned here, even though they are worthy of special recognition.

Frederick Gilbreth's years on the Library Board came during the term of Ralph Phelps, who was more impressed by him than any other board member whom he knew during his twenty-two years as director.[78] Gilbreth was the youngest son of the well-known Gilbreths of *Cheaper by the Dozen* fame. He had followed his illustrious parents, both of whom had been industrial engineers, into an engineering career. He represented the ASME on the Library Board for the period 1953-65, during which time he was vice-chairman (1961-63) and chairman (1963-65), while serving other years on various board committees. Two of his most valuable services occurred in 1958 and 1965, when he made a strong case for more funds for the Library.[79] It is likely he helped bring about the decision of the UET Board to furnish the Library with $90,000, spread over a three-year period. Phelps also recalled that Gilbreth was instrumental in persuading the UET to build the center in New York City. Phelps felt that Gilbreth's modesty kept many of his efforts on behalf of the Library from becoming generally known, just as Gilbreth's mother, Lillian, had played a valuable but unsung role in helping the Library and the center. (For example, she was undoubtedly the one who, through her friendship with President Hoover, persuaded him to take part in the various ceremonies associated with the building of the center.[80]

Frederick Gilbreth exemplified the best of the engineers who, as board members, gave generously of their time and efforts to enhance the status and the well-being of the Engineering Societies Library.

7

The Library Collection

One of the main reasons the Engineering Societies Library has been able to make significant contributions to the engineering profession is the fact that it has such a strong collection. Its scope includes all aspects of engineering and technology, ranging from ablation to zirconium, and its literature represents publishers located in the proverbial four corners of the globe.

Many of the users of the Library are associated with organizations which have their own libraries, a situation which, on the surface, might seem to relegate the Engineering Societies Library to a secondary role. Yet one of its most valuable contributions is to serve as a backup library which shares its technical resources with patrons whenever their regular library cannot meet their needs. Having said this, one must immediately recognize that for another large group of users this library serves as the primary source of technical data.

The backup role which the Library serves for a sizable number of users is proof of the accuracy of the vision of Harrison Craver when, in his ALA presidential address given in 1938, he pointed out the need for developing a number of strong libraries which could provide needed materials for patrons unable to find local sources. As he put it, "The fact that every library can not have a large stock is no reason, though, why unusual needs must be neglected. Ways can certainly be found to surmount this difficulty, by proper planning and cooperation. Behind the first line there is needed a reserve corps of sources for material in slight demand. In other words, we need a system of libraries, and this we do not have as yet."[1] Craver was able to see how his own library had developed over the years into the sort of backup institution he supported, a far cry from the uncoordinated mixture of small collections from four different societies' libraries that he found when his term began.

EARLY STAGES OF THE COLLECTION

When one compares the present condition of the Library collection with the years 1907-17, during which the four Founder Societies moved their separate libraries into their new home, it is readily apparent what great strides have been made. Prior to the move each society had had its own concept of the type of collection to build, ranging from the remarkably overambitious plans minutely described for the ASCE library to the relatively nonexistent goals set for the other three libraries. All depended very heavily on miscellaneous gifts from their members, hardly a reliable way in which to obtain a collection that would meet standards set for it as to scope and size. In view of the more carefully defined plans for the ASCE collection, it is not surprising that by 1900, three years after the society had moved to its attractive headquarters on Fifty-Seventh Street (still standing and serving now as a commercial office building), it had acquired 32,000 volumes of books, pamphlets, maps, photographs, and specifications—a good-sized collection in those days, even if the library had been working at it since 1872.[2] By contrast, the AIEE collection, established around 1885, had only a few hundred volumes by 1900, probably largely due to the society's dependence upon gifts.[3]

The faith the AIEE had in gifts might be said to be justified when one considers the value of the Wheeler gift of some 7,000 books on magnetism and electricity, made to the society in 1901.[4] As previously mentioned, this gift had a special significance aside from its intrinsic value, inasmuch as it figured so heavily in the successful efforts of certain AIEE leaders to obtain from Andrew Carnegie the support and financial aid they sought in order to obtain a new home for their society and its library. Another notable gift was the complete set of the major Italian physics journal, *Nuovo Cimento,* given the library in 1904 by Thomas A. Edison. Part of the set came from his private library and the remainder he purchased; all of it he had uniformly bound in half-morocco.[5]

Exact figures are not available on the size of the combined collections of the three libraries when they moved into the new building in 1907, but it is known that when Miss Howard resigned as director in 1910 the total collection size was less than 45,000 volumes.[6] As Table 5 shows, by the end of 1912 the three libraries had acquired a total of some 46,000 volumes, then added more than 8,400 books in the next three years, bringing the total to approximately 55,000 (before withdrawals were taken into account). Thus the rate of increase—more than 20 percent—in this interval indicated a greater effort towards building the collection.

TABLE 5
GROWTH OF COLLECTIONS, 1912-15[7]

	Volumes	Pamphlets	Totals
Accesssions to 31 Dec. 1912			
ASME	10,289	1,654	11,943
AIEE	16,005	1,448	17,453
AIME	20,237	5,156	25,393
UES	447	90	537
Accessions, 1913-15	8,498	1,897	10,395
Total accessions	55,476	10,245	65,721
Less withdrawals 1913-15			3,275
Total collection 31 Dec. 1915			62,446

It should be noted that a few volumes are listed as belonging to the United Engineering Society. This came about as a result of a decision by the Library Board to pay for items whose contents were of general interest to all three societies, but it was a minor portion of the total. In 1914, for example, the three societies spent a total of $2,666 for their own books while $159 was used to purchase a few UES books.[8] The practice of making each society responsible for buying its own materials was in full agreement with the bylaws adopted in 1912, which said:

> This report [the annual report of the Library Board] shall state the recommendation of the Library Board as to the sum of money which each Founder Society should expend during the coming year for the purchase of books for the library of that society; and as to the sum which the United Engineering Society should expend for the purchase of books or periodicals for the Library of the United Engineering Society....[9]

New procedures and regulations were badly needed if the library collection were to develop in accordance with the overall needs of the UES in mind, rather than to reflect the often divergent views of the three societies. No matter how carefully separate collection policies were prepared on behalf of each of the three societies, it was unlikely such a process could have equalled the quality of a coordinated plan of acquisitions prepared with one unified collection goal in mind.

Beginning in 1914 the Library Board decided to try to improve the sharing of expenses between the societies and the UES, as well as that of ownership of the collection. By 1915 it was agreed by the three societies that the UES would be responsible for all expenditures of the library, including book acquisitions, to be repaid by the societies according to a formula (probably based on the number of members of each). Another important agreement was that while the societies could retain ownership of the books they had originally brought to the joint library as well as any future books they might wish to purchase and place in the Library, the UES would otherwise pay for and own all future purchased items, as well as all gifts given to the Library. This was a crucial decision, for it gave the UES fiscal control of the Library, ended the possibilities of squabbles over who should pay for what, and made a unified acquisition policy much more viable.[10]

These decisions may have spurred the Library on in its efforts to coordinate its separate collections. At any rate, during 1915 work was done on locating duplicates for weeding. There must have been a lot of overlapping of the three collections, because some 16,000 duplicates were identified, mostly periodical volumes.[11]

The addition of the 67,000 volumes brought to the Library when the ASCE moved into the new building in 1917 a few weeks before Craver reported for work increased the total collection size to more than 132,000 volumes, the first time it had exceeded 100,000.[12] There were 22,000 duplicates that the ASCE did not move in, wisely taking the step of identifying them prior to the move and sending them to a group of civil engineers in the Cleveland area.[13] Another development which augured well for the future involved the separate decisions made that year by the four Founder Societies to turn the ownership of their collections over to the UES, with the lead taken by the AIME.[14]

Thus it was more than thirty years before a visionary engineer's dream of a single library acquiring books on a wide range of topics was close to realization, a library which would avoid "the very height of folly that more than one association in the City of New York should undertake to duplicate a series of books; whereas one set of books would answer equally well for all the societies...." These were the words of a man identified only as M. N. Forney, who spoke at the ASCE's annual meeting in 1885.[15] A few others spoke in a similar vein at the meeting, but most of them probably never saw the realization of this goal.

GROWTH OF THE COLLECTION

Craver's arrival did not result in an immediate surge of volumes added to the collection; instead the increase averaged three to four thousand volumes per year. Actually a decrease in size occurred in 1924 when a new method of counting the holdings took effect. The alternatives were to count either physical volumes or distinct bibliographic titles. Up to then the practice had been to count bibliographic titles, but the decision was made to count on the basis of physical volumes. Over the years the Library had put thousands of pamphlets into bound volumes and had been counting each pamphlet as a volume; the new method meant counting each physical volume as one unit regardless of how many pamphlets were bound together within the covers of that volume. The new method was to cause a decrease of tens of thousands of volumes in the total collection count. As Appendix I shows, the total holdings dropped from about 158,000 volumes in 1923 to 110,000 volumes in 1924. It was, however, a decision that put the Library in line with standard practices in most other libraries.[16]

Gifts of books of more than passing importance occurred in 1918 and 1927. The former gift consisted of 8,000 volumes from the General Electric and Westinghouse companies, given after several years of discussion they had had with the library in which they tried (unsuccessfully) to persuade the Library to operate the two company libraries as a branch.[17] The other gift amounted to about 10,000 volumes on geology and mining from a former president of the AIME, Horace V. Winchell.[18] Gifts of this size were rare in subsequent years.

As Appendix I shows, the collection grew in almost linear fashion from 1924 to the present, reaching the total of 255,000 and 10,000 maps by 1980. One interruption of this steady growth occurred in 1960 when a decision was made to withdraw some 18,000 topographic maps because they were already available in the New York Public Library—the decision saved moving them into the new library in the United Engineering Center.[19]

Over the years the Library had consistently collected materials on a wide basis from foreign sources. This was to be of particular value to the government during World War II, when the strength of the German and Japanese collections of the library were put to good use. Some federal agencies regularly sent scores of employees to do their research there. A letter of commendation from the U.S. War Department gave some evidence of the Library's contribution to the war effort.[20] Most uses of foreign material were not so dramatic as this, involving only the more routine aspects of peacetime engineering.

The value of the collection increased steadily over the years, and in 1975 Cabeen employed a retired head of the rare books department of a prominent book store to evaluate the Library's rare books. The estimated value was just over half a million dollars.[21] By 1980 the estimate of the general collection was well over five million dollars, while the rare books had a value of one million dollars.[22] One measure of the richness of its older books may be seen in the fact that the Library was found to own about

49 percent of the 372,000 pages represented in Ralph Shaw's listing of pre-1830 engineering books. Shaw said that he knew of no library which owned as many of the titles as the Engineering Societies Library.[23]

The attitude of the Library towards the possible loss of its rare books was a rather pragmatic one. Realizing that most of them would be difficult, if not impossible, to replace in the event of some disaster, it was decided by the board that there was little point in paying insurance premiums for these volumes. As far as the security of the remainder of the collection was concerned, no check was made on this point until 1931, when an inventory showed that some 1,200 volumes were missing. However, when put on a percentage basis, this amounted to only 0.7 percent of the collection, then totalling 144,000 volumes.[24] Subsequent inventories showed that even during years of greater use of the Library there was little in the way of theft and/or missing volumes. For example, only 400 monographs out of 80,000 were missing in 1955 (or 0.5 percent of the collection), and an inventory in the early 1960s showed that only 450 volumes were missing out of 200,000, a loss of approximately 0.3 percent.[25,26]

The only other aspect of the security of the collection which came to official notice was that of mutilation of books in the old library, some of whose alcoves could not be monitored from the reference desk. Phelps made some changes in shelving practices to put the most likely targets in view of the staff; there is no record of this problem in the new library.[27]

Monographs and Reference Books

No scientific/technical collection would be complete without strong holdings of monographs and books used primarily for reference purposes. The Engineering Societies Library has such strength. One reason it is in this position is due to a fortunate arrangement which developed over many years, beginning around 1917. The annual report for that year stated the Library had received gifts of a number of books from publishers in return for the preparation of book reviews by the library staff, the reviews to be published in various journals of the sponsoring societies. Many of the books were those which could be classified as reference tools, the rest being monographs. It was a mutually advantageous relationship: the publishers benefited by the exposure given their new books in the journals, and over the years the Library saw this small beginning develop into a tremendous boon to its traditionally tight budget for literature. The number of books received gradually increased, until by 1980 the Library was reviewing annually nearly 2,000 books, collectively worth more than $76,000.[28] Thus this activity earned the Library hundreds of thousands of dollars for other purposes, freeing funds that would have been needed to purchase books.

The Library was not totally dependent upon the generosity of publishers for the development of its monograph collection, since there would always be titles that it wanted but did not receive. Normally one would expect the Library to have a simple routine by which the items that were chosen for purchase by the responsible staff member would be ordered by the Acquisitions Department, and indeed this is how it has functioned for some time. But beginning with Cutter's term the Library Board began to intrude in what was essentially and traditionally a prerogative of the head of a library—to be responsible for the selection of materials for the collection. By 1915 the board evolved a complicated plan which called for Cutter to submit lists of proposed books to a board committee for approval, then get bids from dealers, then return to the committee for a final approval on what to buy![29] It was an unworkable and unreasonable plan, not only oblivious to the difficulties involved but also a debasement of Cutter's position as director. No doubt it was due to a general lack of regard the board had for his ability, a situation for which no explanation has been found. Fortunately for the sake of efficiency this ridiculous requirement was lifted after Craver arrived—but not, it should be noted, until four years into his term. A decision made by the board in 1921 to allow him to make purchases without board approval ended a situation which should never have arisen.[30] Since then the library staff members charged with selection responsibilities have been free to select and order whatever they wished.

The reference collection of the Library has always been located close to the desks of the reference staff, both for their convenience and for better security. It consists of encyclopedias, handbooks, directories, dictionaries, indexing services, buyers' guides, and biographical sources, to name some of the more prominent types. Usually only the latest editions are kept in this section, with older editions kept in the stacks. Although all the items are represented in the card catalog, a special directory of these materials was issued in 1970.[31] It was arranged by broad categories, such as civil engineering, mathematics, general reference works, abstract journals, and standards. Although entries were not annotated, full bibliographic citations were supplied. It could serve other librarians as a guide to the types of reference materials an engineering library should have, even if specific titles have been outdated or replaced by now. An update of this bibliography was published in 1979.

Serials

As anyone with experience in technical libraries knows, journals are the heart of the collection. A technical library with only a mediocre collection of monographs could be tolerated as long as its journal holdings were strong. The reason is obvious— journal articles are more current than books and can treat very specific topics in great detail, much more than would be feasible in the average monograph. Therefore it is not surprising that the four Founder Societies libraries forming the Engineering Societies Library were each active in building collections of serial volumes soon after their formation. They did not have much money to spend on serials, so they relied heavily on gifts from members. No doubt there was a great deal of duplication among these libraries, due to the likelihood that their members' gifts were strongest in the major American journals. Evidence of this can be seen from the fact that when the three library collections were being merged in 1915, two-thirds of the 16,000 duplicates were periodical volumes.

However, foreign titles were included, and Edison's gift of a set of the leading Italian physics journal to the AIEE was proof of the scope of their collections. Another clue as to the type of serial collections they maintained can be found in lists they published showing their holdings. One was a separate catalog issued by the AIEE in 1904; at that time it had more than 300 titles, including foreign and domestic journals.[32] Besides the periodicals one might expect the library to have had, there were a few unusual titles, such as the Canadian Society of Civil Engineers *Transactions* and *Horseless Age*. A similar catalog was issued the same year by the AIME library, which had nearly 500 journals.[33] A study of the list shows the greatest strength was in geology and mining journals, many of which were published by state or national governmental agencies.

An excellent picture of the type of serial collection the Engineering Societies Library had around 1915 can be seen in a cooperative list of serials which it prepared on behalf of seven selected libraries in the New York metropolitan area. Earlier union lists had been carried out successfully in the area on a smaller scale. Alice Jane Gates was editor of the 110-page book, which gave the holdings for about 2,000 technical serial titles of the New York Public Library, Columbia University Library, the Chemists' Club Library, the American Society of Civil Engineers Library, the Stevens Institute of Technology Library, and the Plainfield (N.J.) Library, as well as the Library of the Engineering Societies. One of the librarians creadited with having given assistance to the project was Harrison W. Craver, then of the Carnegie Library of Pittsburgh.[34]

A study of the list discloses a great deal about the type of serials collection available at the Library of the Engineering Societies at that time. First of all, it had many foreign titles (such as *Annales des Postes, Télégraphes et Téléphones; Koninklijk Instituut van Ingenieurs. Jaarboekje;* and *Archiv für Electrotechnik*) in the field of engineering besides the English language publications one would expect it to have. Secondly, the Library had a strong collection of titles in the field of pure science, including both English language and foreign titles (such as *South African Journal of Science; American Chemical Society. Journal; Société Géologique de Belgique. Bulletin; Annalen der Physik;* and *Zeitschrift für Physikalische Chemie*). Thirdly, there was, as might have been expected, many titles that the UES Library held alone among the participating libraries (such as *Quarry; Popular Engineer; Electrotechnika (Budapest);* and *American Gas Institute. Gas Institute News*). Fourthly, there was at the same time an unpredictable pattern of duplication of holdings between the UES Library and the other six libraries. It would appear from an informal check that the New York Public Library held the most titles duplicated by other libraries, with Columbia University next in order. The listing of titles held by the library of the American Society of Civil Engineers was a good guide as to what it would some day be bringing to the Library of the Engineering Societies, when it eventually joined the other major societies in the Carnegie building.

There is no evidence that the list was ever used by participants to reduce duplication or to locate areas in which coverage was weak or nonexistent. Fragmented sets were common. The annual report of the Library Board for 1915 cited a published review of the union list which claimed that more than half of the titles were incomplete in each library.[35] The UES Library did try to locate missing issues by sending a want list to libraries as well as to domestic and foreign periodical dealers, but the results were meager. Thus it was evident that the items being sought were not common issues easily located. The next year the board authorized preparation of a supplement to the union list.[36]

Both Cutter and Craver had firm convictions about the importance not only of periodicals but also of indexing and abstracting services for periodicals. Cutter's writings on the subject were quite clear on this point, and Craver's position was evident through his persistent although perhaps ill-advised efforts to create his own periodical index, more or less in competition with *Engineering Index*. Journals obviously rated high with Phelps and Cabeen also, as both kept enlarging the number of titles received, and always met the need of *Engineering Index* for titles it wished to index. Another evidence of their interest in journals occurred during the middle 1960s when Phelps assigned Cabeen, then assistant director, to manage a project using National Science Foundation funds with which to create a computer-based periodicals control system. The sum of $19,000 was awarded in 1966, and though the totally automated system was never realized, by 1968 a computer-based list of current periodicals was pronounced ready for public distribution.[37] Entitled *Periodicals currently received*, it recorded the titles, holdings, call number, and subject scope of each serial. Subsequent updated editions have been issued on a biennial basis, the latest showing receipts as of the close of 1979.[38] Some idea of the geographical scope of the collection can be found by perusing the 256-page publication, which lists approximately 3,000 periodicals (the Library defines a periodical as a publication issued at least four times a year—less than that frequency makes it a "serial" in their terminology). The following titles are typical of the foreign periodicals included: *Geotechnical Engineering* (Thailand), *Bitumen* (Germany), *Acta Polytechnica Scandinavica* (Finland), and *Ingenieria* (Mexico)

The Library's serials collection has benefited over the years from arrangements that brought it many titles either free or on exchange, roughly similar to the system for receiving review copies of monographs. One attraction the Library has had for gift subscriptions is the very strong possibility that many titles would be abstracted by *Engineering Index,* a selection that would enhance the prestige and value of any periodicals so covered. As Table 6 makes clear, during the period from the middle 1950s to the 1960s great progress was made by the Library in improving its serials collection, with a threefold increase made in the number of free journals being received and an overall increase that more than doubled the totals received.

TABLE 6
SOURCES OF SERIALS[39]

Sources	NUMBER OF TITLES	
	1954/55	1960/61
Free	760	2,300
Exchange	290	400
Purchase	370	700
Total titles received	1,420	3,400

By the end of 1979 the Library was receiving more than 7,700 serials (over 5,500 of which were gifts) representing more than fifty countries.[40] It is a strong collection, with most serials covered by *Engineering Index;* this greatly facilitates use of the titles by engineers on a global basis.

Other Materials

One of the strongest features of the collection is a unique set of publications—papers which have been presented at conferences held by the various sponsoring societies, yet for various reasons never published. These papers are known to exist by virtue of listings carried by the appropriate journals of the societies. As far back as 1922 it was agreed that the Library should collect and retain these papers, which it has continued to do.[41] The set now numbers approximately 82,000 items. This strong collection enables the Library to handle requests the various participating societies receive for these papers, a backup service that no other library in the world would be able to perform as well.

Another type of publication that is included in the library's collection, although not so large or valuable a set, is standards. The Library has a very good set of engineering standards from governmental and non-governmental sources. Typical sources include the American Society for Testing and Materials, the British Standards Association, and the German DIN standards. No other technical library in the New York area collects in this field to such an extent.

A very elusive but important type of material is the proceedings of conferences, known also as conference records, transactions, and the like. In recent years the Library has been making a determined effort to increase its holdings of these publications. They often provide the first public announcement of new developments, even predating their appearance in periodical articles. Most of them are covered by *Engineering Index.*

The Library's holdings of technical reports has been very selective. It inherited certain types pertaining to civil engineering that the ASCE library had collected, such as annual reports of municipal waterworks or of railroads, but this sort of report was not pursued in ensuing years due to their lack of technical content. In 1946 the Library acquired a gift of some 33,000 technical reports from the U.S Office of Scientific Research and Development in Washington, which Phelps accepted. Since that time, however, it has not sought nor accepted such governmental reports, partly because of the extensive collection maintained by the Engineering Library at Columbia University, whose holdings currently include more than 800,000 reports from NASA, the AEC and its successor agencies, and the National Technical Information Service.

Likewise the strong holdings of the New York Public Library have negated the need for the Engineering Societies Library to collect another type of literature, namely patents. In 1913 a question was raised at the first meeting of the Library Board regarding the collecting of German patents.[42] A Committee on Acquisition of German patents was created, but it promptly disappeared from the records of the board. Perhaps members of the committee learned of the excellent collection at the New York Public Library and realized the wisdom of relying on that library for patents.

Trade catalogs, or manufacturers' brochures, were once collected in what was a rather haphazard fashion, but they were removed from the collection in 1948, along with about five tons of miscellaneous materials no longer wanted. Before the material was discarded, however, Phelps allowed Columbia University's Engineering Library to make selections of catalogs of a historical nature because it maintained such a collection (which was subsequently transferred to the Smithsonian Institution in Washington).[43]

One type of material which the Library does collect, albeit on a very specific basis, is motion pictures. The collection began to have significance in 1962, when the National Science Foundation signed a contract with the ASME giving the Library $9,700 over a five-year period for the purpose of acquiring technical motion pictures in the field of fluid mechanics. The plan was to encourage the use of the films, to be housed in the Library and loaned as requested.[44] An ASME committee and the Library jointly developed a catalog, put on sale in May of that year, describing the fifty films it already had on hand.[45] This was the first experience the Library had in receiving federal funds for the purpose of developing its collection. A few years later Cabeen coauthored a paper which explained the importance of the Library's scientific film collection and described some of the operational problems it entailed, such as criteria for acceptance, loan regulations, and the methods used for indexing films.[46] The collection now totals nearly a hundred films.

The general collection has always contained maps of all types, including many contributed by the ASCE library. Despite the removal of thousands of topographical maps to avoid continued duplication of the holdings of the New York Public Library, the Library had amassed a total of more than 12,000 maps by the fall of 1980.[47]

8

Cataloging Activities

An essential segment of library activities concerns the cataloging of books so that they can be retrieved by library users and staff. The catalog may not represent every type of material included in a collection, often excluding such items as maps or audiovisual materials. Nevertheless, the catalog remains the primary means for locating books and serials in a library.

BACKGROUND

The catalog of the Engineering Societies Library makes the library rather unusual: it is one of a small number of research libraries in this country which still have a classified catalog, one in which the subject portion is arranged according to a classification scheme rather than by the more usual alphabetical array of subject headings.

The use of classed catalogs has been more common in Europe than in the United States, and the ones that exist in this country have dwindled in number in recent decades. The use of such a type of catalog for a technical library was not unheard of in eastern library circles, as at least two libraries whose histories are intertwined with that of the Engineering Societies Library have had classed catalogs. One was the Department of Technology at Carnegie Library of Pittsburgh, which adopted this type of catalog in 1910 during Craver's directorship.[1] The other

was the library of the American Society of Civil Engineers. Dis-
satisfied with the way the collection was organized, in 1897
Charles Warren Hunt, the secretary-librarian for the ASCE,
developed a classification scheme of his own, which was used in
the creation of a classified card catalog. It took from 1897 until
1899 to reclassify the collection. The next step was the publication in 1900 of a 700-page book catalog of the collection; it listed
approximately 16,000 titles, which represented total holdings of
about 32,000 volumes, pamphlets, maps, photographs, and
specifications. In 1902 a supplement to the book catalog of some
290 pages was published. Evidence of Hunt's interest in the
library classification scheme used for the book catalog and
the card catalog is seen in the fact that he devoted almost fifty
pages to a reproduction of the classification system (as revised
by two members of the library staff) in an article he wrote on the
history of the ASCE.[2]

Another noted technical library which began its existence
with a classed catalog (and has retained it since) is the John
Crerar Library in Chicago, which was created in 1896. The Crerar
catalog was installed upon the decision of its first librarian,
Clement Andrews, who had previously been librarian at MIT.[3]

Special book supplements to card catalogs were published
by various Founder Society libraries in the years prior to their
merger, such as the two-volume book catalog completed by the
AIEE in 1906 which cited and annotated each of the 7,000 or so
titles on electricity and magnetism comprising the famous
Latimer Clark collection, the major gift from Wheeler to the
institute in 1901. As required in the conditions of the gift, an
index was prepared; it was compiled by Brother Potamian, a
professor of physics at Manhattan College in New York City,
who was considered to be an authority on early publications
on these subjects.[4] His manuscript of the index was then reviewed by librarians in several internationally known libraries
before a special AIEE committee published it three years later.
The catalog included, in addition to the indexes and annota-
tions, reproductions of the title pages of selected rare books in
the collection, and a historical account of the purchase and
disposition of the collection.

Other examples of supplemental catalogs were the lists of periodicals published by some of the libraries in the New York area around the turn of the century, as previously mentioned. It is apparent that these were imaginative projects for their times, aimed at making collections more easily used by libraries' primary patrons as well as facilitating their use by other libraries.

THE ENGINEERING SOCIETIES LIBRARY CATALOG

When Cutter arrived at the Library in 1911 he found that each of the three libraries maintained its own catalog, using its own form of the Dewey Decimal System, despite having been "merged" for four years. It was an untenable situation, yet he was unable to convince the board in 1916 of the need to recatalog the collections and establish a single system.[5] The problem was not resolved until two years after Craver had arrived and had made several recommendations regarding the reclassification of the entire collection (by then consisting of four different systems due to the addition of the ASCE collection). However, it was probably complaints from members that forced the board to approve the proposed project. For example, comments received in 1918 from members to the effect that "the Library catalog was so inadequate that they frequently could not obtain the help they needed from it" must have acted as a catalyst to end the stalemate.[6]

Although throughout the records this was usually spoken of as a "reclassification" project, it is clear from other information available that it also involved recataloging, where necessary, to make the different forms of entries found in the four catalogs agree with each other and to create one card catalog. The board approved the allowance of $20,000-$25,000 for the project, specifying that it should start as soon as possible, and that the Library should adopt "the system of classification of the International Institut de Bibliographie as the basis of the classification," as Craver had recommended.[7]

It is not surprising that members complained about the existing catalog records at that time. In a ten-page report on the cataloging problems which he faced and his plan for the project, Craver noted that the existing catalog consisted of: an author catalog made by interfiling cards from the four societies' author catalogs; an alphabetical subject catalog covering three of the four collections; a numerical classed catalog for a subject approach to the ASCE collection, plus an alphabetical index to the classification numbers; and, by now outdated, a printed index to the periodical holdings (the union list published in 1915).[8]

In his report, Craver stated: "It is proposed to substitute for these catalogs an author catalog and a classed subject catalog, the latter to be accompanied by a very full index." In regard to the depth of classification he said:

> we intend to adopt a plan by which the classes on the shelves will be small enough to be conveniently surveyed by a reader, but not so small that extremely long location numbers will be necessary . . . while at the same time the classification in the catalog will be carried to a point where it will be possible to locate a specific subject without examining much related matter.

He pointed out that since the commonly used Dewey Decimal classification was "planned primarily for the arrangement of books, it lacks certain apparatus needed in classed catalogs. It is not sufficiently minute for the latter purpose and it lacks a method for subdividing existing headings in order to express details, various points of view and the relations between different subjects."[9]

Then Craver stated that the Brussels classification, based on an extension of Dewey's scheme, "provides for such needs by the addition of certain symbols to express relations and of detailed general tables of usual relations, which can be added to any subject by means of these marks." He listed some of the features, such as: the sign of addition (+) to indicate that a book treats all the subject numbers connected by it; the colon to indicate the interrelationship of two subjects in the books (like the statistics of steel rolling mills, as represented by 621.77:31, in which 621.77 is the class for rolling mills and 310 the class for statistics); the use of parentheses to indicate form divisions as well as geographical locations; the language sign as indicated by the use of the equal sign (=); and the use of quotation marks to indicate a century in time ("17" would indicate the 1700s, for example). Craver also noted that the published Brussels tables of class numbers consisted of more than twenty-two hundred pages as compared to the eight hundred pages of the Dewey edition of that era, indicating the much more detailed approach possible with the Brussels system.

Even though by April 1919 the full board had not yet ratified the decision of the Executive Committee to go ahead with the project, Craver evidently felt confident enough of the outcome to invite his former employee at the Carnegie Library, Margaret Mann, to come to New York and serve both as head of the cataloging department and as supervisor of the recataloging project. She agreed; the plan was approved by the board the next month.[10]

Not every library with a classified catalog adopted the Universal Decimal Classification, as the Brussels system was sometimes called. Some libraries used the Dewey system without changes. Descriptions of the classified catalogs in such organizations are given in the literature, sometimes with high praise for their benefits and recommendations for their adoption by libraries of all sizes.[11]

An article by Kanardy Taylor noted that the catalog of the Engineering Societies Library was one of "the three outstanding classified catalogs in this country," and named the catalogs at Crerar and Carnegie as the other two.[12] Thus by virtue of Craver's efforts to have a classified catalog established in the Engineering Societies Library, he moved the library into rather exclusive company.

Craver's choice of Mann was a fortunate one, in view of her ultimate success in managing the large project while keeping up with current cataloging duties. As can be seen from Table 7, she and her little staff recataloged a prodigious number of volumes.[13] The project continued for about one year after Mann took a leave of absence in 1924. Cost of the recataloging was approximately $72,000, far above the estimated $25,000 but still less than $1 per volume. The financial burden to the Founder Societies was lessened by the gift of $20,000 from the Carnegie Corpora-

TABLE 7
RECATALOGING PROJECT

Year	Cumulative Number of Volumes Recataloged
1919	6,000
1920	24,000
1921	55,000
1922	69,000
1923	80,000
1924	87,000
1925	95,000

tion available in the period 1921-23; this gift was one measure of the importance with which the project was regarded.[14]

After Mann's departure, the excellence of the catalog was maintained in part due to a low staff turnover in the Cataloging Department. From 1924 until 1965 only two people were in charge of the department: Mary Raymond and Marguerite Soroka. And from 1965 until her retirement in 1980, Soroka kept a close watch on the expansion and updating of the catalog. The catalog Mann created had the traditional parts of a classified system: 1) an alphabetical file of entries for authors or main entry titles; 2) the classified subject file, arranged by class numbers; 3) an alphabetical index to the classified file; 4) a file, arranged by class numbers, of duplicates of the cards found in the alphabetical index. The purpose of the latter file, which was called the control file and was kept in the Cataloging Department, was to enable the classifiers to know what index terms had been assigned to a given class number; if a class number had to be changed or subdivided the classifier could easily tell which index cards in the alphabetical file had to be changed. For example, the index terms assigned to the class 622.338 consisted of "Gas, Natural," "Natural Gas," "Gas Wells," "Oil Wells," and similar phrases. In the control file there would be a card for each index term, filed under the class number 622.338. A description of the system was written in the 1930s by Mary Raymond, for the benefit of those unacquainted with a classified catalog.[15]

Like most library tools, classed catalogs have their ardent proponets as well as their detractors. Library staffs invariably find that the public needs more assistance in learning to use classed catalogs as compared to dictionary catalogs, but at the same time experienced users appreciate the advantages of having all subtopics for a given major topic conveniently arranged in close proximity, eliminating the need for jumping all over the alphabet, as frequently happens in dictionary catalogs. For example, the class for aviation would bring together topics that would range alphabetically from A (ailerons) to L (landing gears) to R (rudders) to W (wings), to cite a few terms. Another advantage is the precise relationships which the experienced user can locate by means of the various symbols built into the system, such as colons, parentheses, and equal signs. Admittedly such niceties may be lost on the casual catalog user; it is safe to say that more use is given the subject section of the classed catalog by the staff and experienced users than by neophytes.

In regard to the subject headings used, most of them are either derived from UDC classes or are based on current usage, the latter being the predominant method for introducing new terms into the system. The cataloging department at various times has experimented with using Library of Congress catalog cards, but there were too many problems to make it worth the effort. For example, delays in delivery of cards were bothersome, and Cataloging in Publication data became common enough to negate much of the value of having cards. Still another reason was that many of the Library's monographs had never been cataloged by the Library of Congress.[16]

FILMING THE CATALOG

Several projects have involved the microfilming of the card catalog, in whole or in part. One arose because of concern for the safety of the catalog in the event of a fire or other source of danger. By 1937 the catalog had grown close to the half-million card mark, with more than 10,000 cards added annually at that time, and its loss would have been a severe blow. So the decision was made to microfilm the entire catalog, with a copy to be stored in a safe place, originally in a bank vault.[17] This was done, and microfilming was continued in later years whenever enough cards had accumulated. In 1955 the reels of microfilm, by then numbering approximately fifty, were moved to a private storage company located outside the metropolitan New York area.[18]

In succeeding years the catalog grew even faster: the rate of card production in 1959/60 of some 12,000 cards per year increased to more than 24,000 in 1967/68 and 33,000 in 1969/70, as Appendix III shows. During the past five years approximately 7,000 volumes and 200 maps per year were cataloged, involving nearly 9,000 subject assignments.[19] A 1962 contract signed by the library resulted in the microfilming of its subject catalog cards in 1963 by the G. K. Hall Company. The selection of the Library was a recognition of the significance of its collection. The filming took nearly three months to complete, preceded by nearly a year of preparation and planning. The resulting catalog occupied thirteen volumes,[20] and from 1964 till 1972 annual supplements were published. In 1975 the annual supplements were superseded by a new annual series, *Bibliographic Guide to Technology,* which combined information from several libraries— it included monographs on techology and engineering cataloged by the Research Libraries of the New York Public Library and the Library of Congress, as well as conference proceedings cataloged by the Engineering Societies Library.[21] A completely new edition of the catalog prepared by G. K. Hall is due for publication near the end of 1981.[22]

The Library's aid in the creation of a new publication was sought in 1971 when a firm known as Research Publications, Inc., sought permission to microfilm all the items listed in Ralph Shaw's bibliography of early engineering books.[23] The Library agreed to permit the filming in return for a copy of the finished complete microfilm record, in the making of which a dozen or so other libraries would participate.[24] However, nothing more could be found in the library records about this project; presumably it was never carried out.

9

User Services

While it is difficult to compare the relative importance of different aspects of the operation of a library, there is little dispute over the importance of the services made available to the users of a library. Many librarians look on this as the ultimate reason for the existence of a library. A library whose only real function is to provide a collection in a passive manner has not made as significant a contribution to its users as it could if it actively sought ways to aid them and then performed these needed services. The Engineering Societies Library has been a far from passive library from the time of its creation, which began with its inheritance of a set of three library staffs (soon four staffs) which had each begun to be relatively active in meeting the informational needs of engineers. This chapter will chronicle the manner in which the Library has provided user services, not only to members of sponsoring societies but also to the general public. Its users have been not only practicing engineers but also engineering students, people in business, lawyers (including patent lawyers), and ordinary laypersons.

THE PREMERGER PERIOD

In view of the years of existence of the separate libraries from which the Engineering Societies Library evolved, it would be useful to trace briefly the types and levels of services these four libraries were offering prior to their ultimate merger in 1907.

Each was created for the primary purpose of serving the members of its parent society, with the approximate dates of origin of the four ranging from 1873 for the American Society of Civil Engineers to 1885 for the American Society of Mechanical Engineers. (It is not easy to determine exactly when these libraries were actually created, due to the lapses of time between the dates of decisions of their boards to organize the libraries and the taking of the first steps to begin them.) Their official charges from their boards were not adequate by modern standards, as can be seen, for example, in the bylaw establishing the ASCE library. It outlined simple clerical duties for the library, such as marking the volumes with ownership stamps. Otherwise the librarian was authorized "to take charge of the Library" and was given no other guidelines as to what sort of services should be offered.[1] The accomplishments of the libraries were largely dependent upon the ambition and skills of their librarians, with counsel given them by their library committees in most cases. No doubt during the early years of these little libraries they were primarily concerned with the mechanics of getting collections started and merely surviving on the small budgets they had at that time. Undoubtedly they soon began to make loans to members and to provide some sort of simple reference service.

By the turn of the century these libraries were able to issue lists of their periodicals for the benefit of users, and their collections were growing larger. With the advent of the merger, they were about to enter an era in which eventually more attention and more funds would be given them, and more would be expected of them. In anticipation of the merger, a survey of ASME members was made in 1906 as to the types of services they would like the new library to perform; the report gives a picture of a library which, for that time, would be considered remarkably active.[2] Somewhere between the origin of the societies' libraries and the period of the move into the new building, the engineers they served had evidently come to expect more than a passive "warehouse" approach to operating a technical library. Note what some of their requests were: performance of research for a fee for those desiring it; circulation of books to members living outside the New York area; provision for a "traveling library" for junior members or for local sections of the society, allowing for long loans; proper indexing of books and journals; notification to members of newly acquired books; and maintaining a collection of patents, lantern slides, and historical information about major engineering projects. Some of these were rather advanced concepts at that time. Some of the ideas were provided for by the new library right away, some took quite a while to bring about, and some were never put into effect.

The Early Period (1907-17)

A very basic matter affecting service should be noted, both because of the early time at which it was established as well as the import it had for the Library. That was the provision in the 1904 charter and bylaws of the UES that "the Board of Trustees shall maintain and conduct a free public library...."[3] This was a departure from the provisions which had governed the individual libraries of the societies prior to the merger, all of which were established to serve only society members. Little is known about the reasons for this significant proviso, although the Wheeler gift did require it to be in a public library. Obviously the new library was also expected to serve members of societies, but the idea of including the public had no clear relationship to planning documents and fund-raising literature in circulation at that time. This stipulation was to affect the nature of the users and the extent of the types of services to be offered from the first day of operation of the Library and continuing throughout its history.

There is little available in the records of the Library to describe the types of services offered users in the period immediately after the initial three society libraries had moved into their new building. It is known, however, that the Library Committee established by the UES did make decisions at that time affecting user services, such as setting the library hours at 9:00 A.M. to 5:00 P.M. on all weekdays, or regulating who would have access to the more valuable books in the reading room (only members of Founder Societies or those introduced by high officials of the societies).[4] Borrowing was not made easy by their regulations—it was allowed only with the written permission of the chairman of the Library Committee, or the secretary of the society which owned the book. However, if there were duplicate copies, they could be loaned at the discretion of the librarian of the society which owned the book. (As can be seen from the latter regulation, all through the early years the concept of a unified library had not yet been grasped and/or accepted.)

Aside from the above-mentioned records, there is little in the way of descriptions about the services provided users in the period from the merger in 1907 until the adoption of the new organization, making the society libraries into a truly unified organization in 1913. That year, however, saw the institution of an annual report and of minutes of board meetings, thus facilitating the tracing of accomplishments.[5] Thus we know that the Library had more than 11,000 personal visitors in 1913. As Appendix II shows, by 1919 the number of personal visitors had doubled.

As early as 1915 there are records regarding the making of photocopies for requesters, with a modest twenty-five hundred copies in 1913, but more than three times that in 1917. One reason for the increase was the acquisition of a library-owned Photostat machine in 1915; prior to then copies had had to be made off the premises. Translation activities were also recorded, beginning with the preparation of seventy translations in 1915 and continuing at about that annual quantity for the rest of this early period. Another way of measuring is to note that by 1917 the Library was translating nearly 350,000 words per year.

One of the prime activities of the Library for many years was performing searches on requested topics, with provision for a fee for all but short searches. The records first show activity in 1914, when nearly two hundred searches were made, reaching approximately five hundred by 1917. Income from searches and photocopies amounted to only $130 the first year, and was never much more than that by 1917. There was no record kept of the types of queries involved in these searches, although many of the searches were added to the collection.

As previously noted, in 1915 the board decided to establish a new section within the Library called the Library Service Bureau, a unit which was to provide search and translation service for a fee for those requiring more than roughly thirty minutes of free assistance.[6] While the Library had been providing service on a fee basis since 1914, the board for unknown reasons placed responsibility for the supervision of the unit with itself, ignoring Cutter's interest in and knowledge of such service. The service was not new—only the name and pattern of organization had changed.

The first year of full operation of the Service Bureau (1916) brought acclaim from the AIME library committee; its report for that year noted that income from searches had increased from approximately $200 in 1915 to some $5,800 in 1916.[7] However, in view of the fact that the expenses of the bureau that year amounted to $5,400, the net profit was actually quite small. Translations totalled eighty-two in 1916. The Service Bureau was the subject of a form letter sent out to potential users in 1915 at the request of the Library Board. One of the services offered was an early version of selective dissemination of information, or current awareness service, as modern terminology would put it. This plan involved the user's mailing in a card indicating the topic(s) on which information was desired; the Service Bureau staff would thereafter send the requestor a card every two weeks, on which was given the author, title, and other bibliographic information for current publications on the topic(s), along with notification of the cost of a copy or a translation of the item noted on the card.[8] There is little evidence of the extent to which this service was utilized.

The rates for paid research had been set by the Library Board at numerous times over the years, with the first mention of this matter occurring in the board's second year of operations. In 1914 it set a rate of $1.00 per hour for searching English language material and $2.00 per hour for foreign languages, with these rates taking effect after thirty minutes of time had been spent.[9] In the spring of 1915 the Executive Committee set the rates for translating work at $2.50 per thousand words of German or

French, and $3.50 per thousand words for other languages. Patent searches were to be charged at the rate of $3.00 per hour.[10]

It is not surprising to learn that in those days of lower labor costs the Library had increased its hours of service, and was normally open from 9:00 A.M. until 10:00 P.M. every weekday (including Saturday) with the exception of certain holidays. Before the start of the 1914 summer season the Executive Committee decided to make one small change—during the summer the Library would close at 6:00 P.M. on Saturdays.[11] The Library could not maintain so many hours of service as the years passed.

THE MIDDLE PERIOD (1918-46)

This period generally matches Craver's term of service. It reflects his interest in providing high-quality user services during a period in which the total number of users served annually (whether personal, telephone, or mail users) increased from a little more than 11,000 in 1918 to more than 45,000 in 1932, as shown in Appendix II. No doubt the depression had much to do with the larger number of patrons in 1932, when unemployment left many people with more time to make use of libraries and, in some cases, to attempt to improve their prospects for employment.

Up until 1918 all statistics for the use of the Library were in terms of personal visits or mail service. After that year a new figure, the number of telephone calls for reference service, was recorded in the Library's annual reports; the impact of the telephone could not be ignored.

At the beginning of Craver's term the most usual means of using the library was through personal visits. But during the period telephone requests increased in number while personal visits slowly decreased. The peak for the number of visitors was reached in 1932/33 at a mark of nearly 32,000, then dipped as low as 20,000 by 1945/46. One reason for the decrease may have been the growing popularity throughout society as a whole of using the telephone in lieu of doing personal errands. Still another reason was undoubtedly the ever-growing geographical area in which the Library's users were located, no longer centering in the greater New York region. Specific figures for telephone usage show an increase from 1,500 in 1918 to approximately 7,000 at the end of the period. It is clear that the nature of user contacts with the Library was changing significantly.[12]

Searches

Except for the first few years the number of inquiries handled by the Library was only recorded if a fee was involved, meaning that it was longer than a thirty-minute search. Some were requested in person and others initiated by a letter or telephone call. The number averaged approximately eighty to ninety per year, with a low point of sixty-four recorded for 1939/40. Earlier years show a higher average, but these records were probably based on a more simplistic definition of a search. A rough analysis of the disciplines involved in the searches shows that they ranked as follows in descending order of frequency: mining and metallurgy, civil and mechanical, although the order of the positions of civil and mechanical searches often switched in this period.

Translations

These were prepared by a few staff members with strong linguistic skills. Phelps recalled that Craver himself used to enjoy preparing translations in between administrative duties.[13] In 1918, 78 were prepared; a peak of 246 (representing well over half a million words) was reached in 1930. While there were variations from year to year as to the most popular languages for which translations were needed, it would appear that the following ranking (in descending order) was typical of most years in this period: German, French, Italian, with fourth place varying between Spanish and Russian.

Loaning of Books

The regulations governing the loaning of books was a topic that repeatedly appeared on board and Executive Committee agendas in this period. For years the rules prohibited loans to anyone. In 1919 the matter arose again, when an employee at the Watertown Arsenal in New England was denied the right to borrow a French periodical from the Engineering Societies Library. His complaint reached the Library Board; at that point, the AIME (to which the employee belonged) was so interested in securing borrowing privileges for its members that it offered five hundred dollars as collateral for any losses involved in loans to its members. The Executive Committee voted to recommend to the full board that loans be permitted under the conditions that the director and a board member approve the loans, and the borrower deposit a sum equal to the value of the book.[14] When the matter went to the full board, the body uncharacteristically rejected the committee's plan. Within a year a slightly more moderate stance was taken when the board agreed that duplicate copies of books could be loaned to members of the Founder Societies, a privilege extended in 1921 to members of other societies which had made financial contributions to the Library.[15]

The rate of five cents per day was set for loans in 1923, when it was also decided that borrowers could buy any duplicate book they had been loaned.[16] In 1927 loan regulations were made a bit more liberal when the board agreed that the director was authorized to make loans to individuals in any unusual circumstances that warranted, in his judgment, granting them.[17] Even with this change, the average member of a society in normal circumstances could borrow only from the relatively small number of duplicate books maintained. The reluctance of the Library Board to make all nonduplicate books available for loan probably stemmed from its desire to ensure that those making personal visits to the Library would never find that a book they wanted was out on loan. However desirable this situation would be for those making personal visits, this attitude would seem to overlook the fact that even then the greatest part of the societies' members did not live close enough to the Library to make such visits conveniently, and were thus denied the privilege of using most of the books available to those able to visit the Library. No one seems to have made a study to determine the likelihood of duplicate requests for the same book occurring within the length of time a loan would have lasted. There were few books loaned in these years, 87 in 1923 and 246 in 1924.

A move that brought the Library into line with general practice in regard to interlibrary loans came in 1924, when the Executive Committee approved the director's making loans to other libraries in North America according to a code recommended by the American Library Association. This code covered such features as what to loan, and length of loans. It was recognized that this new policy would facilitate borrowing by the Engineering Societies Library from other libraries (although there was apparently not much of that done at any time) by establishing good relations with them.[18]

In 1940 the Library Board decided that all volumes in the Library, with the exception of rare or reference books, could be loaned (within the continental United States and Canada) to members of the Founder Societies and to nonprivate and noncommercial libraries. Even bound serials could be loaned. However, nonmembers of the societies could borrow only duplicate books. Several society journals, such as *Mechanical Engineering*, notified their readers of these changes soon after the regulations were adopted, with emphasis on the fact that 150,000 books were now available for loan to members and 10,000 duplicates for nonmembers.[19] These particular regulations had been subject to board discussion since 1937, when a committee first urged a change in the rules. (It should be noted that interlibrary loans were not generally made in succeeding years, despite the provision for them, so individual society members were essentially the only borrowers until the advent of company memberships.)

Following the liberalization of the loan policy in 1940 there was a sudden surge upward in the annual number of volumes loaned per year. Whereas before the changes the highest number had been 380 volumes in 1928, after the new rules went into effect the total jumped to about 1,055 (1942/1943) and to nearly 1,800 a few years after that (1945/1946).[20] Although these figures would seem minuscule in comparison with the circulation records of a college or general public library, by contrast to the early days of the Engineering Societies Library the new totals represented increases on the order of 400 to 500 percent, a dramatic increase for a library serving engineers scattered over a wide area. The response to the new loan regulations showed clearly that the membership had been in need of easier access to to the book collection.

Photocopies

Photocopying at the beginning of the period was at the rate of approximately 6,000 exposures per year; near the end of the period it had increased nearly nine times, reaching a peak of almost 54,000 in 1944/45. As for charges for this service, during the late 1930s the cost of copies was raised to twenty-five cents per page for members of the Founder Societies and thirty cents for all others.[21] A new service was begun in 1939/40 when the production of microfilm for a fee was instituted. Only fourteen films were made in that year, but by 1945/46 the total had increased to eighty-six.

THE LATER PERIOD (1947-80)

This portion of time represents the entire term of Phelps and that of Cabeen up to 1980, during which there was a remarkable increase in the level of service being offered. The changes were evident during Phelps's term when, for the first time, the total number of users went beyond 50,000 in 1958/59. Then, after steady growth, the fiscal year 1978/79 in Cabeen's term marked the initial year that the total number of users exceeded 100,000. This was followed by a still higher level in 1979/80 when there were more than 128,000 patrons, whether by means of personal visits, mail, or telephone contacts. Doubling the number of users in twenty years was ample evidence that the Library's services were needed.

Over the years associate membership had been created; after 1963 it consisted of societies who were permanent occupants of the UES building and gave the same financial support of the Library as Founder Societies. One reason for the upward trend in user statistics can be found in the larger potential number of society members the Library was serving. As Table 8 shows, by 1958 the Associate Societies alone had a total of almost 64,000 members. This, combined with more than 132,000 members of the Founder Societies, gave the Library a potential group of roughly 200,000 engineers to serve. Yet the Library was serving this group of people with a staff smaller than that of 1937, when the members of the Founder Societies numbered no more than 53,000.[22] Directors of the Library have never had more than a bare minimum of staff members with which to work.

TABLE 8

ASSOCIATE SOCIETIES (1958)[23]

Name	Number of Members (Nov. 1957)
American Institute of Consulting Engineers	262
American Institute of Industrial Engineers	5,918
American Society of Heating and Air-Conditioning Engineers	11,415
American Society of Refrigerating Engineers	6,893
American Water Works Administration	12,190
American Welding Society	12,743
Illuminating Engineers Society	8,730
Society of Motion Picture and Television Engineers	5,699
Total	63,850

As time went on there was a continual study of and concern for the hours the Library would be open to the public. Financial conditions triggered a survey of users in the spring of 1950, made over a three-week period.[24] It was found that on the average about seventy-two readers came into the Library on Saturdays from 10:00 A.M. to 6 P.M., quite similar to the weekday average of about seventy-five persons for the same time period. In addition the evening hours on weekdays brought an average of seventeen users per evening from 6:00 P.M. to 10:00 P.M. Later, closing hours were changed to 9:00 P.M. weekdays and 5:00 P.M. on Saturdays. Summer hours became 9:00 A.M. to

5:00 with the Library closed on Saturdays. The subject came up again in 1973 when another survey was made, particularly of evening and Saturday hours. The UET decision then was to close the Library on Saturdays year-round. While the purpose of the move—to save the costs of opening the entire building just for a few Library users—was understandable, it was still considered a drastic step. To offset it at least partially, the Library Board asked that the Library be opened an hour earlier on weekdays.[25] The present schedule was adopted at that time—9:00 A.M. to 7:00 P.M. Monday through Thursday, and 9:00 A.M. to 5:00 P.M. Fridays. No doubt the Saturday closing inconvenienced several users, but no strong protests were received. As will be seen, the number of personal visitors rose to a new high after these changes, so they did not have any obvious permanent effect on the use of the Library.

During the early part of this period there were some 23,000 personal visitors to the Library, which figure was never exceeded until 1968/69, when the total began to increase again, rising to a peak of almost 30,000 visitors by the year ending in the fall of 1980. (This nearly equalled the depression years when 33,000 visitors were recorded). Telephone usage increased from approximately 7,400 in 1946/47 to the highest total ever experienced by the Library, more than 34,000 calls in 1979/80. By this time four separate telephone lines and six instruments were required in the reference area. Calls are automatically counted, increasing the accuracy of the figures and also relieving librarians of a bothersome counting task.[26]

Searches

The number of searches in this period rose and fell, with a high of 168 in 1959/60, preceded and followed by years in which the total fell as low as 38 (1976/77). However, the addition of a computer terminal in 1975 and the subsequent capability of searching databases from the Lockheed Information Service has been slowly building up the number of searches performed. When inquiries are received, reference librarians determine from the nature of the query the likelihood that the callers or patrons may wish to consider a computerized search. If this seems to be the case, they are then referred to the staff member who makes such searches. Sometimes a mixture of manual and computerized searches are made. Taking one month in the fall of 1980 as a sample, nine searches were completed, of which only one was completely done by manual means, the rest being all done with the computer (three) or mixed sources (five). Three other searches were in progress at the end of the month. It is interesting to note that of the twelve requests only two were made in person, the others having been made by telephone (six) or by mail (four). The twelve requestors of the searches represented five industrial companies, one government agency, two law firms, two social service or medical organizations, and two individuals. Topics ranged from illumination to precious metals. This obviously small sample nevertheless gives several interesting indications of the nature of the service—one is the small number of personal requestors, and the other is the varied backgrounds of those requesting a computerized search. Experience so far has shown that the extent and complexity of the questions searched make it difficult to predict accurately in advance the ultimate cost, but limits set in advance by the requestors are adhered to. As an encouragement to individual research, members of Founder and Associate Societies who are making searches for their personal use receive a discount of 20 percent. Searches cost $25 per hour, plus the cost of on-line connection time. Total costs in the sample month ranged from $80 to $617.[27]

Translations

For many years the translation service consisted of finding free-lance translators to whom work could be sent, with staff members such as John Soroka examining and polishing the translations prior to delivering them to requestors. A high point of 306 translations (560,000 words) was reached in 1962/63. It was later decided that the time required was not worth it, and the service was gradually phased out in the late 1970s. Soroka's retirement in 1980 marked the last of any staff members capable of editing translations.

Loaning of Books

During the course of 1953 it was decided that loan regulations for periodicals needed revision. For many years bound volumes had been loaned to qualified requestors. There were many problems, such as wear and tear on the volumes, items unavailable in the Library on certain occasions, and some losses of volumes. The relatively inexpensive new photocopying techniques available made loaning of periodical volumes even less justifiable. The board decided in 1953 that henceforth journals would not be loaned, a decision in line with practice at most other public and academic libraries.[28]

Loans of books amounted only to some 1,600 volumes in 1953/54 and rose to a high mark of nearly 4,400 in 1966/67. In 1979/80 the total was nearly 3,700 volumes; by the end of 1980 there were approximately 20 books being loaned per day. In view of this nearly threefold increase, it is clear that the ban on loaning of periodical volumes had little or no effect on the totals of loans.

Photocopies

During this period some new equipment was introduced, beginning with the long-overdue replacement of a twenty-six-year-old Photostat machine in 1950. More modern copiers were acquired regularly during the ensuing years.

Tremendous changes occurred in the level of copying activity during this period. The lowest number, 43,000 pages made in 1949/50, was a tiny amount compared to later figures. By 1961/62 the 100,000 mark was surpassed for the first time, and only five years passed before the Library was copying more than 200,000 pages per year (1966/67). During the next ten years the number doubled, with the mark of 425,000 set in 1976-77. The following year the half-million point was surpassed, and the 1979/80 figures showed the total as 613,000 copies. This phase of service has become a major source of income for the Library; $400,000 was received in 1978 from photocopy orders.[29] Adding to the funds from staff-made copies were the ones made by patrons on copiers in the reading room, beginning in 1970; each copy cost twenty-five cents. The response was so good that another machine was added, with a third one installed in 1980. During the last year of the period (1979/80) there were almost 165,000 copies made on the reading room machines.

During May 1978 a survey was made of the source of photocopy orders, representing nearly 12,000 pages copied. The results gave very clear evidence of the broad geographical base of users of this service; it was far from being dominated by New York City customers since they constituted only 3.2 percent of the users. Table 9 gives the details of the study.

TABLE 9
SURVEY OF PHOTOCOPY USERS[30]

Percentage	Received from
21.4	New York (exclusive of New York City), New Jersey, Pennsylvania
17.1	New England
12.2	Mexico and overseas
11.5	Ohio, Indiana, Michigan, Kentucky
8.7	Washington, D.C., Virginia, Delaware, the Carolinas, Maryland, West Virginia
6.8	California, Oregon, Washington
6.0	Florida, Georgia, Tennessee, Mississippi, Alabama
5.4	Texas, Oklahoma, Arkansas, Louisiana
3.2	New York City
2.5	Colorado, Arizona, New Mexico, Utah, Nevada, Wyoming, Idaho
2.1	Illinois, Missouri, Nebraska, Kansas
2.0	Canada
1.1	Wisconsin, Minnesota, Iowa, the Dakotas, Montana

In recent years a rush service for obtaining copies was instituted. By paying a five-dollar surcharge, a user is guaranteed that the copies will be mailed within twenty-four hours of receipt of the order. During the survey it was found that approximately 11 percent of the 1,179 orders involved rush service.

Microfilm copies are still being made, although the level of activity at the end of the period (229 orders in 1979/80) had fallen from the peak reached in the late 1960s when 1,000 orders were filled per year.

Bibliography Series

The compilation of bibliographies on a number of subjects of potential interest to a large number of users became a part of the duties of the reference and/or search staff late in the 1940s. Phelps had received so many inquiries from engineers or managers of engineering offices about how the individual engineer could organize a small collection of information that it was decided to compile a bibliography on this topic. This led to the publication in 1948 of what was to be the first of a series of bibliographies prepared by the staff, a compilation of articles on filing and indexing systems. It was to be revised and reissued from time to time after the first printing, the last edition being issued in 1966.[31] By 1961 eight bibliographies were available, dealing with such topics as flow of granular materials, shell molding, adhesive bonding of metals, and a variety of technical subjects. The last one issued was number seventeen, published in 1970. At one point in the 1950s annual sales reached a high of 781 copies.

Company Memberships

In order to facilitate use of the Library by companies and other organizations, a special company membership was established in the early 1970s. By payment of an annual fee (currently one hundred dollars) a company membership is granted, thereby allowing designated people in the company to request loans of books, microfilm copies, or photocopies. Loans of books are charged at the rate of one book for one week equalling one unit; a company membership consists of a hundred units, to be extended upon exhaustion of credit by payment of another hundred dollars. Loans and copying orders may be made by company members by telephone, written orders, or personal visits. Deposit accounts are also available to simplify payments for orders, a service presumably designated for those who order frequently. At the close of 1980 there were more than sixty company members and over thirteen hundred deposit accounts.[32]

10

Networking and Cooperative Activities

It is theoretically possible for a library to isolate itself from other libraries and to avoid participating in the burgeoning cooperative activities which have become prominent in the last decade or so, and still function as a useful library. But most library directors have turned to networks and other cooperative activities with a full appreciation of the potential good these joint enterprises can bring to participants; as library directors, they are painfully aware of the inadequacies of their own libraries and look upon these new ventures as ways for solving some of their local problems. The idea of cooperation among libraries is by no means new—for example, Library of Congress catalog cards have been available since the turn of the century, and no sane librarian would try to devise a different size of card now. Countless other examples could be cited of the long history of cooperative projects among libraries. This chapter will discuss how the Engineering Societies Library has participated in these activities.

COOPERATIVE COLLECTION DEVELOPMENT

One of the basic reasons for the creation of the Engineering Societies Library was the strong conviction among engineers near the end of the nineteenth century that the individual libraries of the various engineering societies then in existence had much in common, particularly the types of materials they collected, and that the libraries would profit from merging. A typical statement for the backers of this idea was the previously mentioned remark of the ASCE member in 1885, that it would be folly to duplicate collections. Yet no matter how logical this argument was, it took nearly a quarter of a century before the Engineering Societies Library was founded. It is ironic that the Library was formed without the participation of the society at which the above remarks were spoken.

This union of four libraries was not the first in the New York City area, but it was the first successful effort ofr engineering libraries in the region. The plans of the Scientific Alliance in the 1890s would have created a joint library to be shared by eight scientific associations located in the city, and would have provided for a collection without the duplication inherent in eight separate libraries.[2]

Other local efforts at cooperative collections included publications of various union lists of serials, which enable libraries to locate publications they lacked. The efforts of the AIEE, AIME, and the UES union lists previously mentioned gave proof of the desire of librarians of that time to engage in cooperative projects.[3,4,5] John Shaw Billings, then director of the New York Public Library, had recommended to Cutter in 1911 that the latter begin the union list project that was to include not only Cutter's library and the New York Public Library but also that of the Chemists' Club, the ASCE, and the Stevens Institute, among others. Thus this successful publication got its start thanks to Billings's foresight and encouragement.

Billings's interest in the Engineering Societies Library dated back at least to the important AIEE dinner in 1903, when he spoke in support of the proposed engineering library because it would serve a different clientele than his and would have certain publications in its collection that his library would not.[6] Evidently he anticipated no serious overlap of the collections.

The consolidation of the society libraries produced some 38,000 duplicate volumes by the time the ASCE library moved in, and this indicated that the merger would reduce duplication before many years passed. There is no way to calculate the exact amount of duplication that was avoided in the years since the founding of the library to the present, but it is undoubtedly in the tens of thousands of volumes.

Another opportunity for a cooperative approach to collections occurred in Phelps's term when he was invited in 1947 by Ralph Beals, then director of the New York Public Library, to discuss how to increase cooperation between the two libraries in acquisitions and collection practices.[7] No progress was made, however.

In 1948 Phelps chaired a meeting of representatives from local technical libraries, the first of several such gatherings to be held that year to discuss cooperative acquisitions. Again, no significant action was taken.[8] It was clear that cooperation was not easily attained.

Still another attempt took place in 1950, again at the invitation of Ralph Beals. Once more Phelps reported to the board that it had been a pleasant but unproductive meeting; the discussion, like the meeting held three years previously, had included the status and role of the Chemists' Club Library.[9] Phelps wanted to find feasible ways of cooperating with local libraries; but since formal mechanisms did not seem viable, he relied on less formal means. Throughout his tenure he maintained friendly relations with the heads of other metropolitan New York libraries, and was able to make informal agreements in cases where they were needed. For example, if the Engineering Societies Library was considering cancelling a subscription to a little-used journal, Phelps would ascertain whether or not it was held by such libraries as Columbia University or the New York Public Library; if so, he would probably rely on their copy and cancel his subscription. Admittedly there was the danger that the remaining library holding a subscription might also cancel it, but Phelps was not too concerned about this.[10]

On numerous occasions Cabeen also kept alive the concept of cooperation in collection development. For example, when budgetary restrictions in the early 1970s caused the Library to reexamine its list of periodicals to see if any titles could be safely discontinued, he contacted Columbia University regarding translated Russian journals. Learning that the titles would be held there, he was able to cancel a number of subscriptions and thus avoid needless duplication.[11] Likewise he was able to assure Columbia librarians that his Library would continue to collect certain national standards, such as those from Great Britain and Germany, allowing other libraries to rely on his for such publications.

Besides local efforts, the directors of the Library had all been aware of national developments. For example, in 1947 Phelps discussed with the Library Board the implications of the Farmington Plan, which called for its participants to agree to acquire foreign materials involving agreed-upon subjects and sources in order to achieve better coverage of foreign literature in this country. It was decided that the Library would not take part, mainly because of the likelihood that the costs for the Library could go as high as $10,000 per year, more than was considered desirable.[12]

REGIONAL AFFILIATIONS

One of the earliest considerations of arrangements whereby the Library could play a formal role outside the New York City area occurred in 1936 when Julian A. Sohon, Craver's principal assistant, was sent on a two-week trip to New England to see if there were some way in which the Engineering Societies Library could aid public libraries in that region.[13] The survey was made as a result of a recommendation from the Library Board's ad hoc Committee on Library Development. Sohon visited twenty-five libraries in Connecticut, Massachusetts, and Rhode Island, and interviewed forty-six engineers about library service. He made several recommendations upon his return: make it easier for other libraries to borrow books (the report did not specify how this was to be done); prepare lists of new books of interest; and arrange occasional visits by staff members from the Engineering Societies Library to give them advice. The board approved the idea of facilitating loans, but rejected staff visits and book lists as being too expensive. Since the question of interlibrary loans had already been more or less taken care of in 1924, when such loans were authorized if done in accordance with ALA rules, the survey seems to have accomplished little.

It now seems rather odd that the Library Board would have suddenly decided to venture into a regional plan of this sort at that particular time. The records are of little help in ascertaining the reasons for this survey. At least it was in line with Craver's conviction of the need to bring library service to regions inadequately served.

The next regional affiliation of any significance was made in 1967 when the library joined METRO.[14] METRO is the unofficial name of the New York Metropolitan Reference and Research Library Agency, a state-supported consortium of libraries in the metropolitan New York area which was formed to foster cooperative activities among such organizations, whether they were public, academic, research, or special libraries. Its membership at that time was small, but soon included the region's major university and public libraries as well as a number of smaller academic and public libraries and several special libraries. Its projects have included a cooperative acquisitions program (whereby expensive reference tools were bought with donated funds and placed in appropriate libraries for the use of all METRO libraries), creation of a special "courtesy" card for use among METRO libraries, and provision for certain types of reference service. Preparation of a computer-based analysis of the collection strengths of participating METRO libraries was carried out in the 1970s, but the Engineering Societies Library did not take part because it felt its strong subject areas were quite obvious. Nevertheless the Library's membership in METRO, which had grown to 105 members by 1980, has brought it into contact with a significant number of libraries in the area and their cooperative projects. Cabeen was active on METRO committees, serving for years as chairman of the Public Services Committee.

Another project with regional implications involved the Library's supplying data on its holdings for a union list of serials published by the New York State Library.[15] The list was issued in 1970, showing the holdings of fifteen libraries throughout the state.[16]

NETWORKING

In a paper on the nature of resource sharing, John Fetterman listed four criteria he felt properly defined an information network.[17] They may be summarized as follows:

1. There must be a formal organization of the participants.
2. There must be rapid communication among participants.
3. Information must move in either direction.
4. There must be some sort of directory of participants to identify the one best equipped to answer a request.

In view of these and similar criteria it is accurate to say that the Engineering Societies Library is currently involved in two networks, one dating back to 1967 and the other back no further than the fall of 1980. Both are serving to increase the resource-sharing activities of the Library and to help it meet the informational needs of engineers. The older alliance is the Library's participation in the interlibrary loan network in New York State, while the newer one involves the establishment of the Library as a source from which users of certain computerized databases can obtain copies of documents.

New York State Interlibrary Loan System (NYSILL)

Early in 1967 the Engineering Societies Library was offered a contract to serve as one of twelve libraries in New York State that would participate as backup or reserve libraries in the NYSILL interlibrary loan program being established.[18] The Library was to receive requests by teletype for engineering materials which could not be filled by the New York State Library; requests would be transmitted from the computer in Albany to whichever library was the most likely source for fulfilling a particular request. There was an initial six-month contract, to be followed by yearly contracts thereafter. The Library Board approved, and work was soon under way.

The plan called for the State Library to reimburse the referral libraries with a participatory grant plus a specified amount for each request handled, with an extra allowance for each item supplied to the original requestor. The current rate is $2.50 for each item searched and an additional $3.50 allowed for each item supplied. Items can either be loaned or photocopied, and are sent directly to the requesting library. An article by Cabeen describing the system noted that in the first year of operation a total of 46,000 requests were handled by the entire system, which is currently operating at a level of about 140,000 requests per year.[19] Statistics for his library for 1979 showed that it filled nearly 4,000 requests out of a total of about 6,400 received. Photocopies were used to fill all but 341 which were for loans of books. The Library is currently averaging about 30 requests per day (or approximately 7,800 per year); the majority of the requests are handled the day they are received, with a reply sent to the Albany computer via teletype the following day. The Library has received commendations for its speed of operation and the high percentage of requests successfully filled. The project has been an important source of income for the Library: in 1979/80 it earned $40,000, bringing the cumulative total over thirteen years to more than $265,000.[20]

This would seem to exemplify a successful blend of modern computers and teletype equipment with traditional interlibrary loan practices, a system which is resource sharing at its best in view of the fact that the tiniest, most remote library in New York State now has access to a huge reservoir of materials. It is also convincing evidence that Craver was very accurate in his 1938 estimate of what was needed to bring about a viable plan for aiding small libraries lacking materials requested of them. His recommendations, dating nearly thirty years before the NYSILL plan was established, were as follows:

> Within a state, those libraries that derive their support from the public purse, through appropriations or direct taxation, could be interrelated, it would seem, into a system of some kind through which every citizen could obtain equal service, regardless of residence. The state library or some large library within the state

might be made a reservoir to which all public libraries
could turn for unusual needs, and be put in a position
to satisfy them, by suitable appropriation from the
public funds.[21]

He favored state-level systems for the first years of operation,
fearing that a national system might be too large for effectiveness.

In comparing the NYSILL plan with the OCLC system,
Cabeen's article pointed out that many libraries do not have
access to OCLC data, some of the referral libraries (including
his) did not normally make interlibrary loans, and some referral
libraries would normally make substantial charges to borrowing
libraries for sending photocopies. So all in all he felt the NYSILL
system was the better of the two plans.[22]

Lockheed Information Systems—DIALORDER Service

In the fall of 1980 the Library completed negotiations with
the Lockheed Information Systems, a major vendor providing
access for the general public to many computerized databases,
whereby the Engineering Societies Library would become affiliated with the Lockheed DIALORDER service. The Library
began serving as one of a select number of organizations which
acted as a paid source of copies of periodical articles, books,
or other items cited in Lockheed's various databases, particularly in COMPENDEX, the computerized version of *Engineering
Index.* Lockheed's customers were notified of the new source in
the fall, and soon requests were received. Customers working at
their computer terminals with Lockheed databases needed
merely to indicate the code name for the Engineering Societies
Library after identifying items in their searches for which they
wanted copies, and the computer would do the rest. It stored up
the customers' names and locations as well as their document
requests, all of which were printed out automatically the following morning on the Library's terminal.

There has been an average of about twenty requests per day,
but a few times it has gone as high as fifty.[23] So far most of the
requests have been for periodical articles and conference papers.
Orders are mailed within three working days of receipt, with a

special rush service available providing twenty-four-hour service for an extra fee of five dollars. The current cost is thirty cents per page plus a five-dollar handling charge per item. If royalty payments are required because of copyrights, these are added on to the other costs.

ENGINEERING SOCIETIES MONOGRAPHS SERIES

This chapter would not be complete without mention of the role of the Library in furthering the publication of an important series of monographs on various engineering topics. It dates back to 1930 when the McGraw-Hill Book Company established the Engineering Monographs Committee, to be chaired by the director of the Engineering Societies Library. The committee's function was to review manuscripts of possible additions to the series, with a royalty paid the library from sales of those monographs which were published.[24] The first royalty check was received in 1932, amounting to some $90.[25] The committee's activity diminished in later years, only to be revived by Phelps in 1947.[26] By the fall of 1980 the series had done so well that the cumulative income to the Library amounted to more than $68,000.[27] It was a strong series— many internationally known engineers authored some of the twenty-three titles. This project has been of real assistance to the funding of the Library, but more importantly it has been a practical example of cooperation between libraries and publishers.

11

Relationship to *Engineering Index*

It is generally accepted that *Engineering Index* is the world's leading abstracting service for the field of engineering. There are several reasons for its prominent position. One is its extensive coverage of all aspects of engineering, unlike some indexes which are confined to a specific discipline. A second reason is its thorough coverage of the literature on an international basis, and a third is the variety of formats in which it is now available—the printed index, the back volumes on microfilm, and a computerized version (COMPENDEX). A fourth reason in many users' eyes is its relationship to the Engineering Societies Library, which maintains copies of all of the many periodical articles, books, conference papers, and reports that are covered by the *Index*—the only exception being a small number of abstracts of government reports indexed in recent years which are not kept by the Library.

The purpose of this chapter is to trace the development of this special relationship and to show how it affects the operations of both organizations. The Library's involvement dates back to its earliest years; there has been a significant impact on it due to its association with this major service, which at present indexes some 2,400 serials and publishes more than 95,000 abstracts per year, featuring a worldwide set of engineering publications.

THE EARLY YEARS (1884-1917)

Engineering Index began its existence on a modest scale in

October 1884 in the form of monthly lists of current periodical articles appearing as a feature of the *Journal of the Association of Engineering Societies,* the section being entitled "Index Notes." At that time it covered only about a hundred engineering and technical periodicals. The association decided, after several years, to issue a cumulated volume; the first of these covered the period 1884-91, which in effect was the first bound volume of *Engineering Index.* (The title actually used was *Descriptive Index of Current Engineering Literature.*) In subsequent years other cumulations were issued for the years 1892-95, 1896-1900, and 1901-05, after which it cumulated on an annual basis. By 1905 the index was citing about ten thousand articles per year. The monthly sections were moved to a different journal, *The Engineering Magazine* (later entitled *Industrial Management*), for the period 1896-1918. By 1918 the coverage included about 250 periodicals.[1]

Cutter showed his interest as early as 1914 in the Library's indexing of engineering periodicals when the annual report noted that he had had the library staff start a card index for articles in certain journals. In 1911 he had written an article which pointed out the need for better coverage of the world's technical periodical literature by indexes. He stressed the need for inclusion of many foreign language journals, not limiting indexes to English language materials, and he recommended that the subject portion be arranged by a classification scheme, with an additional author and subject index. Comments prepared by reviewers were printed with his article; reactions ranged from general support to outright opposition. Several letters referred to *Engineering Index,* an indexing service that had been in existence for several years; the comments about it included a recommendation by one writer that the Engineering Societies should sponsor it, while another person felt its foreign coverage should be strengthened.[2]

The *Index* had been well established by the time Cutter's article appeared and his comments had been recorded in the 1914 annual report. Obviously the coverage of the *Index* was not broad enough to satisfy him or several persons who commented on the matter after his article was published. Cutter himself brought up the topic again in the annual report for 1915; he pointed out the H. W. Wilson Company's *Industrial Arts Index* indexed only 74 titles and *Engineering Index* only 218, with very little foreign coverage. His criteria for a suitable index included weekly issues, semiannual cumulations, and a classified section (to aid foreign users not able to handle an alphabetical index in English as easily as a classed index).[3] Just as his recommendations for funds for recataloging the collection brought no response, so did his comments on periodical indexes go unnoticed officially.

THE MIDDLE YEARS (1917-46)

Beginning with the term of Craver, the affairs of *Engineering Index* and the Library grew progressively more intertwined. In October 1917 the managing editor of the *Index*, L. P. Alford, offered to sell his publication to the UES. It would have been an arrangement pleasing to Craver, as he had long hoped for a better index to technical periodicals, just as Cutter had; and it would have meant the *Index* would have strong backing (from the UES). However, the Library Board rejected the offer despite Craver's urgings to the contrary.[4]

Within six months another proposal involving the indexing of periodicals came along. This time it involved the Canadian Mining Institute, which recommended that an index to periodicals on mining be prepared by a joint United States-Canadian team of engineering societies. The Executive Committee studies the idea during meetings in the spring of 1918. No direct participation was authorized; instead the committee approved a study "to consider possible methods of centralizing and coordinating the collection and distribution of industrial information and to report its recommendations. . . ."[5] Craver had recommended a more general study, not restricted to "industrial information," but at least he had been able to interest the board in an analysis of technical information and the problems of gathering and using it. More was heard about the Canadian society in June when the AIME announced that it wanted a monthly periodical index on mining topics and that the Canadian Mining Institute wanted to help.

Only a few months later discussions with the AIME over the disposition of a $100,000 gift from James Douglas reached a point at which the Executive Committee announced its recommendation that part of the annual income from the gift ($3,000) be used for the preparation of an index to mining and metallurgical periodicals.[6] It is not difficult to imagine that Craver played a major role in the committee's decision to make this recommendation. Although the final agreement on the use of the Douglas money was not clearly stated in the records, the annual report showed that Craver had started a monthly index on mining and related topics and stated he hoped it would be a forerunner of a more general index to all disciplines in engineering. About 10 percent of the coverage was of foreign journals.[7]

A major development was announced in 1919: the ASME bought *Engineering Index* and continued to produce a multidisciplinary index.[8] The Library began to cooperate in several ways, such as allowing the *Index* staff to borrow library copies of journals and books to index, thus saving the staff the cost of purchasing the literature itself. It now seems odd that the Library made no charge for this service at that time and did not do so until decades later, a situation that may have been due in part to Craver's great enthusiasm for the production of a good technical periodical index as well as to the fact that *Engineering Index* in the early years was far from a profitable venture, so that cutting costs was vital to its existence.

In the annual report for 1921 there is further word on indexing activities. It stated that the Library "assists the American Society of Mechanical Engineers to compile the Engineering Index" and also "prepares the Mining and Metallurgical Index published by the American Institute of Mining and Metallurgical Engineers." The AIME's index, which the Library prepared, appeared in the AIME *Bulletin*. At that time the American Society of Civil Engineers issued its own index to the literature of its field. Craver continued to hope that all these efforts could be combined in one publication, but the Executive Committee rejected this idea at a meeting in 1924.[9]

Evidence of other indexing projects is found in the annual report for 1924. Besides regularly preparing the indexes for "Management Engineering" and "Mining and Metallurgy," the Library was also indexing "Explosives Engineer." In addition it began a small card index service, probably on an experimental basis, for twenty-four subscribers, who were sent a total of 5,400 cards during that year.[10] Little more was written about the experiment after that year.

In 1925 the Library Board was asked to consider the merger of *Engineering Index* with the *Mining & Metallurgical Index;* the board referred the matter to the Executive Committee,[11] but there is no record of any action taken by the committee on this question. At the previous meeting the committee had been asked to consider sponsoring the preparation of a card index service for periodical articles that would be available on a fee basis to subscribers. No action was taken.[12] Thus the Library Board rejected several opportunities to unit the Library with *Engineering Index.* The importance of a periodical index was stressed by Craver at a board meeting in 1927, at which time it was estimated that $5,000 would be required for the Library to start such a project. The board declined to add any money to the budget for such a project until a more definite plan could be presented.[13]

As an indication of the ways in which ideas about indexing seemed to be interchanged between the library staff and the staff of *Engineering Index,* it is interesting to note that in 1928 *Engineering Index* began to provide the very same sort of card service to subscribers that the Library Board's Executive Committee had rejected as a library service when Craver proposed the idea in 1925.[14] The cards listed the same information found in the bound indexes and were mailed to subscribers to the various subject sections available. This enabled much quicker dissemination of information that the conventional plan, still in force, of printing a listing of citations in a section in a journal each month.

As a result of the continuing and increasing amount of interplay between the Library and *Engineering Index,* it is not surprising that in 1928 the Library Board was invited by the board of *Engineering Index* to appoint one person (plus an alternate) to serve on the advisory board for *Engineering Index.* Within the next few months the Library Board decided its own chairman would be the logical one to be so appointed, and he was subsequently elected chairman of the advisory board, with Craver named as the alternate representative. This was to be only the first of a long succession of appointments from the Library on various boards or in various offices of *Engineering Index*.[15]

During 1929 Craver found a source of money that enabled him to begin a relatively large-scale project for indexing periodical articles. The unexpected funds came in the form of the income from a $50,000 gift to the Library made by James H. McGraw. Once McGraw's intention to make a gift was known, Craver spoke to him in the spring of the year about the need for such an index, particularly one with a broader coverage of journals than *Engineering Index* provided. Later Craver wrote a letter to Earl Whitehorne, an associate of McGraw at the McGraw-Hill Publishing Company; Craver proposed a budget for the project of $10,000 per year, a sum he estimated as adequate to cover the cost of hiring two indexers and two clerks.[16] McGraw gave his approval to Craver's proposed index project, and the board added its blessings to the plan, stating:

> The need for a comprehensive index to these periodicals has been recognized for years by every one who has had occasion to consult them. It has been remarked in your [Craver's] reports from their commencement, as well as elsewhere. The valuable index issued by The American Society of Mechanical Engineers is ameliorating the situation, but it has not attempted until recently to cover the whole range of engineering. The guides to past periodicals are distressingly inadequate.... A card index to these publications is planned, which will bring together in a uniform arrangement the material now scattered in hundreds of volumes, and complete it by filling the gaps that exist at present. Mr. McGraw's gift will enable you [Craver] to undertake this work....[17]

By the end of the following year Craver announced that some 33,000 entries had been made in the new index. The total had grown to more than 100,000 items by the end of 1933.[18] Craver obviously wasted no time in his eagerness to index periodicals. Ralph Shaw wrote that Craver hoped in time to index not only current journal articles but also to work backwards to 1900, an ambitious plan for an index being prepared by a handful of people, some of them working on a part-time basis.[19]

As late as 1933 the Library had another opportunity to pur-; chase *Engineering Index* when the ASME proposed such a sale, no doubt prompted by the fact that the *Index* had a $20,000 deficit by that time. Again, the board refused to become financially involved with *Engineering Index*.[20] While the board continued to reject such opportunities, Craver pushed ahead, once the McGraw money became available, with his own index, even it it meant duplicating the goals, if not the exact coverage, of *Engineering Index*.

After the Library Board declined to purchase *Engineering Index* in 1933, its publisher, the American Society of Mechanical Engineers, continued to search for a solution to its financial burden. Sometime during 1934 the answer was found close at hand. As Russell Shank described it, ownership was shifted from the ASME to

> a nonprofit corporation organized by the employees of the service and a few members of the American Society of Mechanical Engineers. The Society still owned the title to *Engineering Index* and the publication's offices remained in the same building with the Society's headquarters and the Engineering Societies Library, whose serials it used in the abstracting operations.

But, as Shank pointed out, "none of the other societies had been induced to assist in its publication."[21]

The reluctance of the Library Board to take on this sort of financial involvement carried over to the fall of 1934 when Craver's name had been proposed by officials of *Engineering Index* as a full member (unlike his previous alternate status) of the board of the publication. However, the Library Board, perhaps due to the financial problems of *Engineering Index* had been experiencing, declined to allow him to take the office because of possible financial involvement of the Library if the *Index*'s funding problems continued. Matters dragged on until, in 1940, the board officially refused to allow its members to serve on the *Engineering Index* Board under any conditions.[22] Despite this conservative step, one that was reversed later, there was no change in the long-term cooperation at the operating level on a day-to-day basis by the staffs of the Library and the *Index*.

In looking back at Craver's term, it now seems strange that none of the Library Boards with whom he worked ever questioned his devotion to creating his own periodical index, so obviously in competition with *Engineering Index,* and thus a waste of time and funds.

THE LATER YEARS (1947-80)

Phelps had a good appreciation of the importance of maintaining close relationships between Library and *Engineering Index.* There were several reasons for this position. A major one was the benefits that came to the library through having its collection virtually all covered by *Engineering Index*. This meant that a large worldwide group of users would be aware of the details of most of the Library's holdings; it also meant his own staff would be able to retrieve information more quickly due to this coverage. Consequently Phelps did all that he could to promote close cooperation between his staff and that of the *Index*. It was done willingly, in recognition of the indispensable role that the *Index* was playing in the sci-tech library world.

It is not surprising that Phelps was willing to devote so much of his time serving on both the Board of Trustees and the Board of Directors of the *Index*, a relationship which began in 1955 and continued until 1974[23]. It is most likely both organizations profited from his participation.

Financial problems were given particular attention at a 1964 Library Board meeting, and various remedies were proposed at the time. *Engineering Index* was also having its fiscal troubles, so it may not have been a great surprise for Library Board members to learn later that the UET Board had created an Ad Hoc Library-Index Merger Committee to consider the possibilities of a merger of the two entities. The Library Board asked its Publicity Committee to discuss this possibility from the standpoint of fund raising. The ad hoc committee reported at the Library Board meeting in 1965 that there were at least no legal obstacles to such a move. In order to familiarize Library Board members with both organizations, there was an inspection tour of both facilities.[24] However, no further action was taken in this matter.

In the mid-1960s the Library and *Engineering Index* both became the subject of discussion for a confusing array of committees and study groups, some of which overlapped with others. For example, the Bowie Committee (named after a board member, R. M. Bowie) succeeded in obtaining a grant of $35,000 from the Engineering Foundation in 1964 for studying the informational needs of engineers. One of its recommendations was that the *Index* and the Library be combined in a proposed nationwide engineering system to be known as the United Engineering Information Center. Little was accomplished by the committee, but one practical result was the National Science Foundation grant which Cabeen supervised for the production of the library's machine-readable records for its serials, as previously mentioned.[25]

By 1965 still another study group, known as the Tripartite Committee, was created for devising a national engineering system and center.[26] All through its plans was the expectation that both *Engineering Index* and the Library would play a major role in the rather nebulous national plan being evolved, but after long and expensive study, the Battelle Memorial Institute proposed no role for either the Library or the *Index*.[27] The response to the Battelle report was essentially negative.

The Larsen Study proposed in 1970 that the Library seek an NSF grant to study the role of the Library and *Engineering Index* in connection with automation of information systems; nothing come of this plan.[28] Almost immediately after came the North Study. Its report, issued in 1971, attempted to identify possible future roles for the Library, some of which, such as critical abstracting and current awareness service, were seen by evaluators as already being done by *Engineering Index*.[29] Again the results were minimal, if not nonexistent. The spate of studies of the Library and *Engineering Index* ended at about that time, although the boards of the two organizations have continued informal analyses of the direction in which they should go in the years ahead.

Cabeen continued the tradition of close contact with *Engineering Index* in many ways, one important example being his twelve years of service as a member of its Board of Directors; furthermore, he served as secretary and as treasurer during some of his years on that board. There was no question that the two organizations have worked closely together during his term, and the long-standing routines of the Library, such as providing literature for indexing or coordinating acquisitions to meet the needs of the *Index,* were continued. An important event in the relationship occurred in the early 1970s when *Engineering Index* began to make financial contributions to the Library, giving as much as $35,000 in some years.[30] The aid given the *Index* by the Library was finally being reimbursed to some extent.

One of the few indications of a possible overlapping of functions of the two organizations came about late in 1980 when the *Index* publicly announced the establishment of what it called the Engineering Information Search Service. It issued a brochure describing searches that could be done on a fee basis, using either *Engineering Index* or the facilities of the Library. Bibliographies, abstracts, or answers to specific questions were listed as types of service to be offered. There was no mention made of the rates to be charged, although informal sources indicated a minimum figure of seventy-five dollars was actually set. It remains to be seen if this service will become popular and what effect, if any, it will have on the Library's reference service.

12

A Look at the Future

There will naturally be circumstances in the future of the Engineering Societies Library which cannot possibly be foreseen at this time. On the other hand, it is possible to identify some basic principles which might very well be meaningful in the future. The goal of this brief chapter is to present broad concepts on topics of more than passing importance to the library.

COLLECTIONS

There is very little to criticize in regard to the Library's collection policies. As will be seen, the general tone of the statements in this section is that of urging continuation of present activities.

It is important that the Library continue to work closely with *Engineering Index* in the selection of materials needed for indexing. The advantages to the Library of having its collection indexed by an internationally known set of index tools (printed, microfilmed, and computerized) need no restatement here. It is a unique arrangement that both organizations are fortunate to share.

The Library must continue to keep in close contact with libraries in the area to see that full advantage is taken of the strengths of each in planning collections. The Library at present concentrates on the discipline of engineering in all its aspects. There may be more benefit to other libraries in examining their policies in relation to what the Engineering Societies Library is doing than the latter is apt to gain.

The Library must be ready to increase its capability to give reference service using computerized databases. The rapid acceptance of this service in the few months since it has been featured more in contacts with users is probably a good indicator of future activities. A gradual increase in staff time and/or equipment may be necessary for the Library to respond promptly to demand for this service.

The number of potential users for the Library is very impressive when one considers the hundreds of thousands of members of the Founder Societies alone, much less the number of members of Associate Societies or unaffiliated engineers. Yet there are probably many engineers who are still largely unaware of the valuable collection and services the Library has to offer. Every dollar spent in publicizing the strengths of the Library to potential users would surely be worthwhile, if a reasonable plan is developed and if consistent effort is made. It is essential that a way be found to keep the name of the Library fresh in the minds of potential users. As new engineers are constantly entering the profession, only regular attention to the need for publicity will serve the purpose.

Close cooperation must be maintained with *Engineering Index* so that it and the Library complement each other's activities in serving users. Competition along these lines would not be productive in the long run. In general, this has not been a problem, but the potential for it is always present.

It would be wise to keep a close watch on the need of library users for technical translations. At one time the Library was translating over half a million words per year. The linguistic skills of engineers has probably diminished rather than improved in recent years, and the need for translations may be as great as ever. Granted that obtaining staff members capable of making accurate translations was never an easy task, different ways of providing this service, if it is still needed, may have to be developed, such as using part-time translators. For example, New York City has a large number of engineering students, many at the graduate level, who have a background in various foreign languages. Many experienced engineers with linguistic skills also are available in the area.

TECHNICAL SERVICES

The classed catalogue makes the cataloging needs of the Library very different from that of libraries using conventional subject cataloging systems. However, careful attention should be paid to improvements in the transmission of descriptive cataloging data; perhaps ways will be found which would result in the lowering of cataloging costs as far as the descriptive portion is concerned, even if original cataloging seems necessary for the subject portion of the process.

NETWORKING

This section is also largely concerned with urging continued efforts along lines already taken, in view of what the Library has accomplished to date.

The sharing of the Library's resources to the greatest extent feasible is important to its welfare as well as to that of the general technical information sector. Networks are one important means of accomplishing this sharing, and ways now being used by the Library seem most suitable. Networks need not be joined with the expectation that financial losses will occur; on the contrary, the present activities are proving to be useful sources of income. The combining of networking with photocopying, in itself a useful and profitable activity of the Library, seems to hold the most promise. Every opportunity along these lines must be sought and given careful consideration.

APPENDIX I

COLLECTION SIZE

Thousands
of volumes

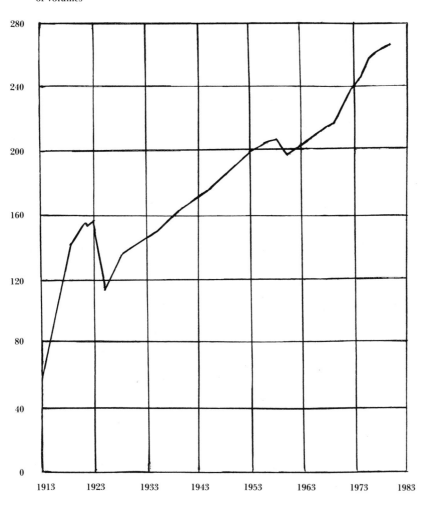

175

APPENDIX II

TOTAL NUMBER OF USERS; TELEPHONE USERS

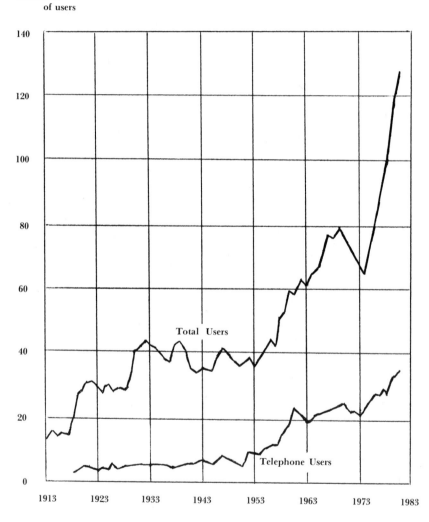

Thousands
of users

Total Users

Telephone Users

1913   1923   1933   1943   1953   1963   1973   1983

176

APPENDIX III

NUMBER OF TITLES CATALOGED;
CATALOG CARDS ADDED

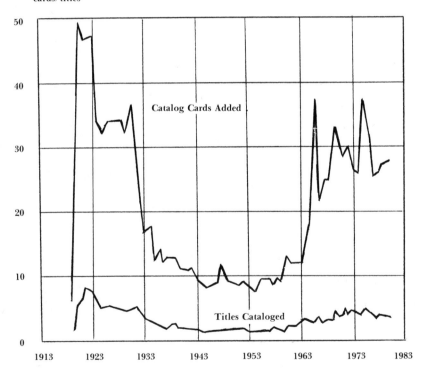

Thousands of
cards/titles

Notes

Preface

1. Mount, Ellis. *History of the Engineering Societies Library, 1913-1973.* New York: Columbia University, School of Library Service; 1979. 376 p. Dissertation.

Chapter 1

1. Shaw, Ralph R. *Engineering books available in America prior to 1830.* New York: Columbia University, School of Library Service; 1931. Master's essay.
2. ESL. Library Board. Executive Committee. *Minutes.* 1915 Jan. 6.
3. ESL. Library Board. *Annual Report.* 1915.
4. ASCE. Board of Direction. *Minutes.* 1916 Oct. 10; 1917 Apr. 17-18.
5. ESL. Library Board. *Annual Report.* 1979/80.

Chapter 2

1. Rae, John B. The invention of invention. *In:* Kranzberg, Melvin, Pursell, Carroll W., Jr., eds. *Technology in western civilization.* New York: Oxford University Press; 1967: p. 328.

2. De Camp, L. Sprague. *The ancient engineers*. New York: Doubleday; 1963: p. 26.

3. Kirby, Richard Shelton [and others]. *Engineering in history.* New York: McGraw-Hill; 1956: p. 24-25.

4. Haupt, Lewis M., ed. *American engineering register.* New York: Engineering News Publishing Co.; 1885.

5. Armytage, W. H. G. *A social history of engineering.* 4th ed. Boulder, CO: Westview Press; 1976: p. 179.

6. Fubini, Eugene G. The need for action. *In:* Chen, Kan, ed. *Technology and social institutions.* New York: Institute of Electrical and Electronic Engineers; 1974: Chap. 1.

7. Branscomb, Lewis M. Engineers in an ionized world. *In:* Chen, Kan, ed. *Technology and social institutions.* New York: Institute of Electrical and Electronic Engineers; 1974: p. 133.

8. Frey, Carl. Engineering: what does the future hold? *Mechanical Engineering.* 93(9): 18-20; 1971 Sept.

9. U.S. Bureau of the Census. *Historical statistics of the United States: colonial times to 1970.* Bicentennial edition. Washington: Government Printing Office; 1975: p. 140.

10. Armytage. *Op. cit.* p. 25, 62, 75-76, 95, 109, 162, 188-189.

11. Rae. *Op. cit.* p. 331-332.

12. Merritt, Raymond H. *Engineering in American society: 1850-1875.* Lexington, KY: University Press of Kentucky; 1969: Chap. 2.

13. Warren, S. Edward. *Notes on polytechnic schools.* New York: Wiley;   1866: p. 6-7.

14. Haupt. *Op. cit.* p. xx.

15. Ross, Earle Dudley. *Democracy's college: the landgrant movement in the formative stage.* Ames, IA: Iowa State College Press; 1942: p. 155, 156.

16. Frey. *Op. cit.* p. 20.

17. Emmerson, George. *Engineering education: a social history.* New York: Crane, Russet; 1973.

18. Thornton, John L.; Tully, R.I.J. *Scientific books, libraries and collectors.* 3d ed. London: Library Association; 1971: Chap. 1-2.

19. Houghton, Bernard. *Scientific periodicals—their historical development, characteristics and control.* Hamden, CT: Linnet; 1975: Chap. 1.

20. Armytage. *Op. cit.* p. 77-78.

21. *Ibid.* p. 100-105.

22. *Ibid.* p. 122-123.

23. *Ibid.* p. 130-131.

24. *Ibid.* p. 358.

25. Wisely, William H., *The American civil engineer, 1852-1974; the history, traditions and development of the American Society of Civil Engineers.* New York: American Society of Civil Engineers; 1974: p. 6-13.

26. Hunt, Charles Warren. The first fifty years of the American Society of Civil Engineers. ASCE. *Transactions*. 48: 220-226; 1902.

27. Wisely. *Op. cit.* p. 17.

28. Hunt. *Op. cit.* p. 224.

29. Proceedings of meetings, May, 1871 to February, 1873. Wilkes-Barre Meeting, May 16, 1871. AIME. *Transactions*. 1:3-5; 1871-73.

30. Acting Secretary's report. ASME. *Transactions*. 1:3-4; 1880.

31. Hutton, Frederick Remsen. *A history of the American Society of Mechanical Engineers from 1880 to 1915*. New York: American Society of Mechanical Engineers; 1915: p. 16-17.

32. Rules of the American Society of Mechanical Engineers. ASME. *Transactions*. 2:xx; 1881.

33. Sinclair, Bruce, with the assistance of James P. Hull. *A centennial history of the American Society of Mechanical Engineers, 1880-1980*. Toronto: University of Toronto Press; 1980.

34. Historical preface. AIEE. *Transactions*. 1:1-4; 1884.

35. Rules of American Institute of Electrical Engineers. AIEE. *Transactions*. 1 (appendix): 1-5; 1884.

Chapter 3

1. Tornow, William H. *The Engineering Societies Library: a history of its origins and early development, 1852-1928*. Brookville, NY: Long Island University, Graduate Library School; 1966. Master's essay.

2. American Society of Civil Engineers. *Constitution, by-laws and act of incorporation*. New York: the Society; 1867: p. 4-5.

3. On founding an engineering library and museum. ASCE. *Proceedings*. 1:225-227; 1875.

4. Tornow. *Op. cit.* p. 39, 45.

5. Wisely, William H. *The American civil engineer, 1852-1974; the history, tradition and development of the American Society of Civil Engineers*. New York: American Society of Civil Engineers; 1974: p. 42-43.

6. Raymond, Esther. An old engineering library. *Library Journal*. 39(4): 285-286; 1914 Apr.

7. The twenty-eighth annual meeting of the Society, November 3d, 17th and 18th, 1880. ASCE. *Proceedings*. 6:106; 1880.

8. Hunt, Charles Warren. The activities of the American Society of Civil Engineers during the past twenty-five years. ASCE. *Transactions*. 82:1595; 1917.

9. Tornow. *Op. cit.* p. 45-47.

10. Hunt. *Op. cit.* p. 1572-1652.

11. Coxe, Eckley B. [and others]. Centennial Committee of the American Institute of Mining Engineers. AIME. *Transactions*. 4:12; 1876.

12. Proceedings of meetings, 1877. AIME. *Transactions. 5:24-25, 32-37; 1877.*

13. Raymond, R.W. Biographical notice of Thomas M. Drown, M.D., LL. D. AIME. *Transactions.* 36:296; 1906.

14. Annual report of the Council, AIME. *Transactions.* 30:xxiii; 1900.

15. By-laws. AIME. 36:xxvi-xxvii; 1906.

16. Report of the Library Committee. AIME. *Bi-monthly Bulletin.* (8):xvii; 1906 March.

17. Martin, T.C. Annual Meeting. Report of the Secretary. AIEE. *Transactions.* 2:2-3, 14; 1885 May 19-20.

18. Weaver, William D., ed. *Catalogue of the Wheeler gift of books, pamphlets and periodicals in the library of the American Institute of Electrical Engineers, with introduction, descriptive and critical notes by Brother Potamian.* New York: American Institute of Electrical Engineers; 1909; p. 467.

19. Hutton, Frederick Remsen. *A history of the American Society of Mechanical Engineers from 1880 to 1915.* New York: American Society of Mechanical Engineers; 1915: p. 267-268.

20. Towne, Henry R. Report of the Committee on Library (sic). ASME. *Transactions.* 6:11-13; 1884-85.

21. Hutton. *Op. cit.* p. 269-274.

22. Lydenberg, Harry Miller. *History of the New York Public Library, Astor, Lenox and Tilden Foundations.* New York: New York Public Library; 1923. Boston: Gregg Press; 1972: Chap. 14.

23. Dain, Phyllis. *The New York Public Library: a history of its founding and early years.* New York: New York Public Library; 1972: p. 2.

24. Lydenberg. *Op. cit.* Chap. 1, 2.

25. Dain, *Op. cit.* p. 43.

26. Bolton, H. Carrington. A plea for a library of science in New York City. *In:* Council of the Scientific Alliance of New York. *Addresses delivered at the first joint meeting held at the American Museum of Natural History, Tuesday Evening, November 15, 1892.* New York: [the Alliance]; 1893: p. 42-46.

27. Cole, George Watson. *Early library development in New York State (1800-1900).* New York: New York Public Library; 1927: p. 19.

28. Council of the Scientific Alliance. Builing Committee. *Report.* New York: [the Alliance]; 1898.

29. Gamble, William W. Technology and Patent Division of the New York Public Library. *Library Journal.* 36(12): 634-635; 1911 Dec.

30. New York Public Library. *Annual Report.* 1911. p. 131.

31. Dain. *Op. cit.* p. 328.

32. *Ibid.* p. 210-212, 214-216.

33. *Ibid.* p. 267-268.

34. General Society of Mechanics and Tradesmen. *Historical sketch and government.* [New York: the Society]; 1914.

35. Cooper Union for the Advancement of Science and Art. *Annual Report.* 5:7; 1864.

36. Cooper Union. *Annual Report.* 50:49-54; 1909.

37. Tauber, Maurice; Cook, C. Donald; Logsdon, Richard H. *The Columbia University Libraries: a report on present and future needs prepared for the President's Committee on the Educational Future of the University.* New York: Columbia University Press; 1958: p. 10-12.

38. *Catalogue of technical periodicals: libraries in the City of New York and vicinity.* Edited by Alice Jane Gates. New York: Library of the Engineering Societies; 1915: p. xii.

39. Periodicals relating to chemistry and physics in the New York Public Library and Columbia University Library. New York Public *Bulletin.* 1(6): 152-158; 1897 June.

40. Pratt Institute Free Library. *Annual Report.* 18: 8-11; 1905.

41. Kruzas, Anthony Thomas. *Business and industrial libraries in the United States, 1820-1940.* New York: Special Libraries Association; 1965: Chap. 3.

42. *Ibid.* Chap. 4.

43. *Ibid.* p. 8.

44. *Catalogue of technical periodicals. Op. cit.* p. xii.

45. Bay, J. Christian. *The John Crerar Library, 1895-1944: an historical report prepared under the authority of the Board of Directors.* Chicago: John Crerar Library; 1945: Chap. 2.

46. *Ibid.* p. 29-47.

47. *Ibid.* p. 29.

48. *Ibid.* p. 65, 113.

49. *Ibid.* Chap. 5, 10.

50. Keeping pace with the times. *Carnegie Magazine.* 44(1):5-14; 1970 Jan.

51. *Carnegie Institute and Carnegie Library of Pittsburgh.* 6th ed. Pittsburgh; Carnegie Institute and Carnegie Library of Pittsburgh; 1930.

52. Bobinski, George S. *Carnegie libraries: their history and impact on American public library development.* Chicago: American Library Association; 1969: p. 11-14.

53. Carnegie Library of Pittsburgh. *Annual Report.* 1; 1897.

54. Carnegie Library of Pittsburgh. *Annual Report.* 6; 1902.

55. Kruzas. *Op. cit.* p. 35-36.

56. Merritt, Raymond H. *Engineering in American Society: 1850-1875.* Lexington, KY: University of Kentucky Press; 1969: p. 10-11.

Chapter 4

1. The Annual Meeting, January 21, 1885. ASCE. *Proceedings*. 11: 21-23; 1885.

2. *Ibid.* p. 24-27.

3. Martin, T. C. Annual Meeting. Report of the Secretary. AIEE. *Transactions*. 2:2-3; 1885 May 19-20.

4. *Ibid.* p. 14.

5. House Fund of the American Institute of Electrical Engineers. *Electrical World*. 9(24): 276; 1887 June 11.

6. The Annual Meeting of the Society, January 18th and 19th, 1888. ASCE. *Proceedings*. 14:2-3; 1888.

7. Weaver, William D. [*Letter to Andrew Carnegie*]. 1895 Nov. Photocopy located at: Engineering Societies Library, New York.

8. Finnegan, P. W. [*Letter to W. D. Weaver*]. 1896 Dec. 4. Photocopy located at: Engineering Societies Library, New York.

9. Weaver, William D. [*Miscellaneous correspondence.*] Photocopy located at: Engineering Societies Library, New York.

10. Obituary notices—Josiah Latimer Clark. Institution of Electrical Engineers. *Proceedings*. 28:671; 1899.

11. Weaver, William D., ed. *Catalogue of the Wheeler gift of books, pamphlets and periodicals in the library of the American Institute of Electrical Engineers, with introduction, descriptive and critical notes by Brother Potamian*. New York: American Institute of Electrical Engineers; 1909: p. vi-vii, 468.

12. Dunn, Gano. [*Letter to John H. R. Arms.*] 1948 Jan. 13. Photocopy located at: Engineering Societies Library, New York.

13. Steinmetz, C. P. [*Letter to Andrew Carnegie.*] 1901 Dec. 2. Photocopy located at: Engineering Societies Library, New York.

14. Scott, Charles F. The Institute's first half century. *Electrical Engineering*. 53(5):659; 1934 May.

15. Scott, Charles F. The engineeer of the twentieth century. AIEE. *Transactions*. 20:305; 1902.

16. Scott. *The Institute's first half century*. p. 660.

17. Library dinner of the American Institute of Electrical Engineers. AIEE. *Transactions*. 21:104-105; 1903.

18. *Ibid.* p. 117, 119.

19. Andrew Carnegie speaks to electrical engineers. *New York Times*. 1903 Feb. 10:9 (col. 3).

20. Scott. *The Institute's first half century*. p. 660.

21. The Carnegie gift to engineering. ASME. *Transactions*. 24:870; 1903.

22. Scott. *The Institute's first half century*. p. 660.

23. Engineers' Club purchase. *New York Times*. 1903 Feb. 7:16 (col. 1).

24. United Engineering Trustees. *History, charter and by-laws of United Engineering Trustees.* New York: the Trustees; 1965: p. 8.

25. National clubhouse for Engineers. *New York Times.* 1903 May 4:3 (col. 3).

26. Annual Report of the Council. AIME. *Transactions.* 35:xxxiv; 1905.

27. Scott. *The Institute's first half century.* p. 660.

28. Carnegie offer refused. *New York Times.* 1904 Mar. 3:1 (col. 6).

29. Hunt, Charles Warren. The activities of the American Society of Civil Engineers during the past twenty-five years. ASCE. *Transactions.* 82:1595; 1917.

30. AIME. Council. *Minutes of meetings.* 1904 Mar. 10.

31. Scott. *The Institute's first half century.* p. 661.

32. Carnegie, Andrew. [*Letter to Gentlemen of The Mechanical Engineers, Institute of Mining Engineers, Institute Electrical Engineers, Engineers' Club New York City*] 1904 March 14. Located at: Library of Congress, Manuscript Division, Washington, DC; Carnegie Papers.

33. United Engineering Trustees. *Op. cit.* p. 8.

34. Andrew Carnegie gift to engineers. *New York Times.* 1904 Mar. 16:1 (col. 3).

35. Plans for Carnegie home for engineers. *New York Times.* 1904 Apr. 24:9 (col. 3).

36. H. D. Hale winning architect. *New York Times.* 1904 July 14:7 (col. 7).

37. United Engineering Trustees. *Op. cit.* p. 13.

38. Carnegie building plans. *New York Times.* 1905 May 14:6 (col. 1).

39. United Engineering Trustees. *Op. cit.* p. 8.

40. Americans good mixers, says Andrew Carnegie. *New York Times.* 1906 May 9: 8 (col. 7).

41. United Engineering Society. *Op. cit.* p. 8-9.

42. [United Engineering Society.] *The Engineering Societies Building.* [New York: the Society]; n.d. p. 3-4, 6-7, 15. Located at: Engineering Societies Library, New York.

43. Engineers open their new home. *New York Times.* 1907 Apr. 17:18 (col. 2).

44. *Ibid.*

45. *Ibid.*

46. United Engineering Society. *Op. cit.* p. 9.

47. Cutter, W. P. The Engineering Societies Library. *Library Journal.* 39(12):880 (photo), 894-897; 1914 Dec.

48. Hutton, Frederick Remsen. *A history of the American Society of Mechanical Engineers from 1880 to 1915.* New York: American Society of Mechnical Engineers; 1915: p. 191.

49. Scott. *The Institute's first half century.* p. 660-661.

50. [United Engineering Society.] *The Engineering Building.* New York: the Society; 1906. Located at: Engineering Societies Library, New York.

Chapter 5

1. United Engineering Trustees. *History, charter and by-laws of United Engineering Trustees.* New York: the Trustees; 1965; p. 8.

2. Annual report of the Council. Report of the Library Committee. ASME. *Transactions.* 28:25-26; 1906.

3. *Who's who in library service.* New York: Wilson; 1933: p. 179.

4. Cutter, William Parker. *Charles Ammi Cutter.* Chicago: American Library Association; 1931. Boston: Gregg Press; 1972: p. 58.

5. Cutter, Charles Ammi. *Rules for a dictionary catalog.* 4th ed. Edited by W. P. Cutter. Washington: Government Printing Office; 1904. London: Library Association; 1935.

6. Necrology. *Library Journal.* 60(2):532; 1935 June 15.

7. Cutter, William P. Rare books and their values. *In:* Dibdin, Thomas Frognall. *The bibliomania, or book-madness: history, symptoms and cure of this fatal disease.* Boston: The Bibliophile Society; 1903: vol. 1.

8. William Parker Cutter. *Agricultural Library Notes.* 10(6):245; 1935 June.

9. Cutter, Charles Ammi. *Dictionary of American library biography.* Littleton, CO: Libraries Unlimited; 1978; p. 109-115.

10. *American Library Annual.* 1911-12: 162.

11. Annual report of the Council. ASME. *Transactions.* 34: 8-9; 1913.

12. Report of the Board of Directors for fiscal year ending April 30, 1913. Library Committee. AIEE. *Transactions.* 32:2176; 1913.

13. United Engineering Society. *Charter and by-laws, 1912.* New York: the Society; 1912.

14. ESL. Library Board. *Minutes.* 1913 Feb. 6.

15. ESL. Library Board. Executive Committee. *Minutes.* 1913 Mar. 4, Apr. 1, June 4.

16. ESL. Library Board. *Minutes.* 1913 May 1, Sept. 22.

17. ESL. Library Board. *Annual Report.* 1916.

18. Cutter, Walter [sic] Parker. An international technical index. *Special Libraries.* 2(8):83-86; 1911 Oct.

19. ESL. Library Board. *Annual Report.* 1914.

20. ESL. Library Board. *Annual Report.* 1915.

21. ESL. Library Board. Executive Committee. *Minutes.* 1915 May 5, May 14, June 10.

Notes 187

22. ESL. Library Board. Executive Committee. *Minutes*. 1914 Sept. 2.

23. ESL. Library Board. Executive Committee. *Minutes*. 1914 Dec. 2.

24. ESL. Library Board. Executive Committee. *Minutes*. 1915 Mar. 3, May 5.

25. ESL. Library Board. Executive Committee. *Minutes*. 1915 May 5, May 14, June 10.

26. ESL. Library Board. Executive Committee. *Minutes*. 1916 Mar. 1.

27. ESL. Library Board. Executive Committee. *Minutes*. 1916 Jan. 9, Apr. 5.

28. ESL. Library Board. Executive Committee. *Minutes*. 1917 Feb. 7.

29. *Who's Who in Library Service. Op. cit.* p. 132.

30. Cutter, W. P. The Technical library's field of service. *Special Libraries*. 6(9): 150-152; 1915 Nov.

31. Cutter, W. P. The Engineering Societies Library. *Library Journal*. 39(12): 880, 894-897; 1914 Dec.

32. Cutter, W. P. Periodicals—as a librarian views them: with some suggestions as to standardization. *Publishers' Weekly*. 86(25): 2038, 2040, 2041; 1914 Dec. 19.

33. ESL. Library Board. *Annual Report*. 1917.

34. Cutter, W. P. A new classification for the Baker Library, Harvard Graduate School of Business Administration. *Special Libraries*. 20(3): 72-74; 1929 March.

35. William Parker Cutter. *Agricultural Library Notes. Op. cit.* p. 245.

36. ASCE. Board of Direction. *Minutes*. 1915 May 5, Sept. 20-21.

37. ASCE. Board of Direction. *Minutes*. 1915 Sept. 20-21.

38. ASCE. Board of Direction. *Minutes*. 1916 June 23-24.

39. ASCE. Board of Direction. *Minutes*. 1916 Oct. 10; 1917 Apr. 17-18.

40. *Interview with John Soroka*. 1978.

41. Carnegie Library of Pittsburgh. *Annual Report*. 1902; 1903.

42. Carnegie Library of Pittsburgh. *Annual Report*. 1907; 1908.

43. Koskoff, David E. *The Mellons: the chronicle of America's richest family*. New York: Crowell; 1978: p. 67, 68.

44. Craver, Harrison W. [*Letter to Andrew Carnegie.*] 1908 Nov. 16. Copy located at: Carnegie Library of Pittsburgh, Pennsylvania Room.

45. Carnegie Library of Pittsburgh. *Annual Report*. 1910.

46. Resignation of Mr. Craver, Carnegie Library of Pittsburgh. *Monthly Bulletin*. 22(3):149-150; 1917 Mar.

47. ESL. Library Board. Executive Committee. *Minutes*. 1917 Feb. 7, Mar. 7, Apr. 4.

48. ESL. Library Board. *Minutes.* 1917 May 10, Oct. 18.
49. ESL. Library Board. *Minutes.* 1917 May 10.
50. ESL. Library Board. Executive Committee. *Minutes.* 1917 May 2; Sept. 5.
51. United Engineering Trustees. *History, charter and by-laws.* 1965. *Op. cit.* p. 9.
52. ESL. Library Board. *Minutes.* 1918 Jan. 10.
53. ESL. Library Board. Executive Committee. *Minutes.* 1918 Feb. 6.
54. ESL. Library Board. *Minutes.* 1924 Jan. 10; 1926 Jan. 14, Oct. 14.
55. ESL. Library Board. *Annual Report.* 1929.
56. ESL. Library Board. *Minutes.* 1925 May 14.
57. ESL. Library Board. *Minutes.* 1928 Oct. 11.
58. United Engineering Trustees. *History, charter and by-laws.* 1965. *Op. cit.* p. 11.
59. ESL. Library Board. *Annual Report.* 1917.
60. ESL. Library Board. *Minutes.* 1917 Oct. 18, Nov. 7.
61. ESL. Library Board. Executive Committee. *Minutes.* 1918 Oct. 2.
62. ESL. Library Board. *Annual Report.* 1921.
63. ESL. Library Board. Executive Committee. *Minutes.* 1925 Mar. 19.
64. Phelps, Ralph H. Engineering. *Library Trends.* 15 (4):869; 1967 Apr.
65. ESL. Library Board. Executive Committee. *Minutes.* 1928 Aug. 9; 1929 Mar. 14, June 20.
66. ESL. Library Board. *Annual Report.* 1929.
67. ESL. Library Board. *Annual Report.* 1917.
68. ESL. Library Board. *Annual Report.* 1926.
69. ESL. Library Board. *Minutes.* 1919 May 8, Oct. 9; 1920 Oct. 14; 1921 Jan. 13.
70. ESL. Library Board. *Minutes.* 1945 Jan. 11.
71. Craver, Harrison W. [and others]. The Farey diary, 1819. Newcomen Society. *Transactions.* 16:215-219; 1937.
72. ESL. Library Board. Executive Committee. *Minutes.* 1937 Sept. 14; 1938 Dec. 12.
73. Craver, Harrison Warwick. *Dictionary of American library biography. Op. cit.* p. 102.
74. Craver, Harrison W. Unfilled business. *Library Journal.* 63(12): 481-485; 1938 June 15.
75. Sullivan, Peggy. *Carl H. Milan and the American Library Association.* New York: H. W. Wilson; 1976: p. 169-173.
76. ESL. Library Board. Executive Committee. *Minutes.* 1939 June 12.

77. Craver, Harrison W. [*Letter to (Theodore W.) Koch*] 1939 May 18. Carbon copy located at: Columbia University, Library Service Library, New York.

78. Craver, Harrison W. The role of the library in engineering education. *In: The inauguration of Oliver C. Carmichael as Chancellor at Vanderbilt University and a symposium on higher education in the South.* Held Feb. 4, 1938. Nashville, TN: Vanderbilt University; 1938: p. 248-256.

79. *To Harrison W. Craver in commemoration of his twenty-five years of distinguished service as Director of the Engineering Societies Library, April 1, 1942.* [New York: United Engineering Trustees]; 1943.

80. *Ibid.*

81. ESL. Library Board. *Minutes.* 1942 Oct. 8.

82. ESL. Library Board. Executive Committee. *Minutes.* 1944 Sept. 20.

83. ESL. Library Board. Executive Committee. *Minutes.* 1945 Apr. 18.

84. ESL. Library Board. *Minutes.* 1946 Jan. 10, May 9, Oct. 17.

85. Shaw, Ralph R. Harrison W. Craver. *Library Journal.* 76(15): 1295; 1951 Sept. 1.

86. Shaw, Ralph R. Personnel. *College & Research Libraries.* 7(4): 347-348; 1946 Oct.

87. Shaw, Ralph R. At the top of the beanstalk. *Library Journal.* 92(9): 1808; 1967 May 1.

88. Gjelsness, Rudolph H. Margaret Mann. *Library Journal.* 86(6); 1123; 1961 Mar. 15.

89. Carnegie Library of Pittsburgh. *Annual Report.* 1907.

90. Bishop, William Warner. Margaret Mann. *In: Catalogers' and classifiers' yearbook.* Chicago: American Library Association; 1938: p. 11-12.

91. Mann, Margaret. *Dictionary of American library biography.* *Op. cit.* p. 339-340.

92. Mann, Margaret. [*Letter to Harrison W. Craver.*] 1942 Mar. 18. Located at: Engineering Societies Library, New York.

93. Grotzinger, Laurel A. Women who "speak for themselves." *College & Research Libraries.* 39(3): 180-181; 1978.

94. Shaw, Ralph R. *Dictionary of American library biography.* *Op. cit.* p. 476-481.

95. Shaw, Ralph R. *Engineering books available in America prior to 1830.* New York: Columbia University, School of Library Service; 1931. Master's essay.

96. *Interview with Mary Raymond.* 1978.

97. ESL. Library Board. *Minutes.* 1937 Oct. 14.

98. Rapid selector patented. *Library Journal.* 78(4): 310; 1953 Feb. 15.

99. McDonough, Roger H. Ralph R. Shaw. *Library Journal.* 97(22): 3952; 1972 Dec. 15.

Chapter 6

1. *Interview with Ralph Phelps.* 1977.
2. *Interviews with various librarians.* 1977-78.
3. *Interview with Ralph Phelps.* 1977.
4. ESL. Library Board. *Minutes.* 1946 Jan. 10, Oct. 17.
5. ESL. Library Board. Executive Committee. *Minutes.* 1945 Dec. 19.
6. ESL. Library Board. Executive Committee. *Minutes.* 1945 Dec. 19; 1946 Feb. 21, June 11, Nov. 19; 1947 Apr. 15.
7. ESL. Library Board. *Annual Report.* 1950/51.
8. ESL. Library Board. *Minutes.* 1948 Oct. 21, Nov. 18; 1949 Mar. 3
9. ESL. Library Board. *Minutes.* 1958 May 13.
10. ESL. Library Board. *Minutes.* 1947 Oct. 16.
11. ESL. Library Board. *Minutes.* 1948 Feb. 24.
12. *Personnel communication from Ralph Phelps.* 1977.
13. ESL. Library Board. *Minutes.* 1947 Oct. 16.
14. *Interview with Ralph Phelps.* 1977.
15. ESL. Library Board. *Minutes.* 1962 May 17, June 21, Sept. 20.
16. ESL. Library Board. *Minutes.* 1965 Feb. 18, Apr. 15, May 20.
17. ESL. Library Board. *Minutes.* 1965 Sept. 16.
18. Tripartite Committee. *Action plan for the establishment of a United Engineering Information Service.* New York: the Committee; 1969: p. 1.
19. Wisely, William H. *The American civil engineer, 1852-1974; the history, traditions and development of the American Society of Civil Engineers.* New York: American Society of Civil Engineers; 1974: p. 423.
20. ESL. Library Board. Executive Committee. *Minutes.* 1937 June 18.
21. United Engineering Trustees. *History, charter and by-laws of United Engineering Trustees.* New York: the Trustees; 1965: p. 10, 14.
22. ESL. Library Board. *Study of space requirements of the Engineering Societies Library.* 1954 Aug. *In:* ESL. Library Board. *Minutes.* 1954 Sept. 21: appendix.
23. ESL. Library Board. *Minutes.* 1955 June 9.
24. ESL. Library Board. *Minutes.* 1956 Nov. 15; 1957 Jan. 17.
25. ESL. Library Board. *Minutes.* 1958 Feb. 13.
26. ESL. Library Board. *Minutes.* 1958 June 19.
27. Delaney, Edmund T. *New York's Turtle Bay old & new.* Barre, MA: Barre Publishers; 1956; p. 45-63.

28. *Interview with Ralph Phelps*. 1977.
29. *Ibid.*
30. Hoover salutes engineer groups. *New York Times*. 1959 Oct. 2: 11 (col. 4).
31. Freedom of mind urged by Hoover. *New York Times*. 1960 June 17: 33 (col. 5).
32. *United Engineering Center*. New York: United Engineering Trustees; [n.d.]. Located at: Engineering Societies Library, New York.
33. Remarks by Mayor Robert F. Wagner at cornerstone laying ceremonies of new United Engineering Building (June 16, 1960). *Municipal Engineers Journal*. 46(2):74-76; 1960.
34. United Engineering Trustees. *Op. cit.* p. 11.
35. Remarks by Mayor Robert F. Wagner at dedication of United Engineering Center Building. *Municipal Engineers Journal.*   47(4): 196-197; 1961.
36. Hoover finds lag in engineers. *New York Times*. 1961 Nov. 10:17 (col. 4).
37. Engineers sell 39th st. building given by Carnegie 56 years ago. *New York Times*. 1960 Oct. 23 (Sec. VIII):1 (col. 3).
38. Phelps, Ralph H. Planning the new library: Engineering Societies Library. *Special Libraries*. 53(5):274-281; 1962 May-June.
39. ESL. Library Board. *Minutes*. 1967 Jan. 19.
40. ESL. Library Board. *Annual Report*. 1960/61.
41. *Interview with Robert Goodrich*. 1978.
42. ESL. Library Board. *Minutes*. 1961 May 18.
43. *Interview with Ralph Phelps*. 1977.
44. ESL. Library Board. *Minutes*. 1963 June 20, Sept. 19; 1964 Jan. 16, Feb. 20.
45. ESL. Library Board. *Minutes*. 1964 June 18, Nov. 19; 1965 Feb. 18, Dec. 9.
46. ESL. Library Board. *Minutes*. 1964 Apr. 16, May 21, June 18, Nov. 19.
47. ESL. Library Board. *Minutes*. 1967. Jan. 19, Feb. 16.
48. *Personal communication from Marguerite Soroka*. 1979.
49. ESL. Library Board. *Annual Report*. 1954/55; 1960/61.
50. ESL. Library Board. *Annual Report*. 1959/60; 1960/61.
51. *Personal communication from Eugene B. Jackson*. 1979.
52. *Personal communication from Ralph Phelps*. 1978.
53. Phelps, R. H. Engineering information—all is not lost. *Mechanical Engineering*. 80(12):68-69; 1958 Dec.; also published as: ASME Paper no. 58-A-178, presented at American Society of Mechanical Engineers Annual Meeting, held in New York on Nov. 30-Dec. 5, 1958; *Science*, 129 (3340): 25-27; 1959 Jan. 2:*In:* Sharp, Harold S., ed. *Readings in special librarianship*. New York: Scarecrow. 1963: p. 416-423.
54. Crane, E. J. [*Letter to Dr. (sic) Ralph H. Phelps*]. 1959 Jan. 20. Located at: Engineering Societies Library, New York.

55. ESL. Library Board. *Minutes.* 1968 Jan. 18, Feb. 15, Apr. 18, May 16, Sept. 19.

56. *Interviews with various employees and colleagues.* 1978.

57. *Interview with S. K. Cabeen.* 1977.

58. ESL. Library Board. *Minutes.* 1964 Jan. 16; 1965 Dec. 9; 1968 May 16.

59. *A biographical directory of librarians in the United States and Canada.* 8th ed. Chicago: American Library Association; 1970: p. 153.

60. ESL. Library Board. *Minutes.* 1972 Jan. 21.

61. ESL. Library Board. *Annual Report.* 1971/72.

62. ESL. Library Board. *Annual Report.* 1979/80.

63. ESL. Library Board. *Minutes.* 1972 Nov. 10.

64. ESL. Library Board. *Annual Report.* 1972/73.

65. ESL. Library Board. *Annual Report.* 1979/80.

66. ESL. Library Board. *Minutes.* 1972 Jan. 21, Apr. 21: appendix.

67. Science librarians face budget bind. *Chemical & Engineering ws.* 50(10): 22, 25; 1972 Mar. 6.

68. ESL. Library Board. *Minutes.* 1969 May 22, Oct. 30.

69. Tripartite Committee. *Action plan for the establishment of a United Engineering Information Service.* New York: the Committee; 1969.

70. Liston, David M., Jr. [and others]. United Engineering Information Systems study. *In:* National Engineering Information Conference. *Proceedings.* Held June 24-25, 1969, Washington, D.C. New York: Engineers Joint Council; [1970]: p. 69-81.

71. *Interview with Ben Weil.* 1978.

72. ESL. Library Board. *Minutes.* 1971 Sept. 17, Oct. 22.

73. ESL. Library Board. *Annual Report.* 1979/80.

74. ESL. Library Board. *Annual Report.* 1972/73; 1979/80.

75. ESL. *Summary of 1,179 photocopy orders received during the month of May 1978.* New York: the Library; [1979].

76. *Interview with Carmela Carbone and Carol Tschudi.* 1980.

77. ESL. Library Board. *Annual Report.* 1967/68; 1979/80.

78. *Interview with Ralph Phelps.* 1977.

79. ESL. Library Board. *Minutes.* 1958 Apr. 17: appendix; 1965 Oct. 21.

80. *Interview with Frederick Gilbreth.* 1978.

Chapter 7

1. Craver, Harrison W. Unfilled business. *Library Journal.* 63(12); 481-485; 1938 June 15 (p. 483).

Notes 193

2. Hunt, Charles Warren. The activity of the American Society of Civil Engineers during the past twenty-five years. ASCE *Trans-actions.* 82:1595; 1917.

3. Weaver, William D., ed. *Catalogue of the Wheeler gift of books, pamphlets and periodicals in the library of the American Institute of Electrical Engineers, with introduction, descriptive and critical notes by Brother Potamian.* New York: American Institute of Electrical Engineers; 1904: p. 467.

4. Weaver. *Op. cit.* p. vi-vii.

5. Annual report of the Council. Report of the Library Committee. ASME. *Transactions.* 28:25-26; 1906.

6. Mount, Ellis. *History of the Engineering Societies Library, 1913-1973.* New York, Columbia University, School of Library Service; 1979: p. 115. Dissertation.

7. *Ibid.*

8. ESL. Library Board. *Annual Report.* 1914.

9. United Engineering Society. *Charter and by-laws, 1912.* New York: the Society; 1912.

10. ESL. Library Board. *Annual Report.* 1915.

11. *Ibid.*

12. ASCE. Board of Direction. *Minutes.* 1916 Oct. 10; 1917 Apr. 17-18.

13. Hunt. *Op. cit.* p. 1593.

14. AIME. Board of Directors. *Minutes.* 1917 Mar. 23.

15. The Annual Meeting, January 21, 1885. ASCE. *Proceedings.* 11: 21-23; 1885.

16. ESL. Library Board. *Annual Report.* 1923, 1924.

17. ESL. Library Board. *Minutes.* 1918 May 9.

18. ESL. Library Board. *Minutes.* 1926 Jan. 14; 1927 Jan. 13.

19. ESL. Library Board. *Annual Report.* 1959/60; 1960/61; 1979/80.

20. ESL. Library Board. *Minutes.* 1945 Jan. 11.

21. ESL. Library Board. *Minutes.* 1975 Nov. 21.

22. ESL. Library Board. *Minutes.* 1980 Nov. 17.

23. ESL. Library Board. *Minutes.* 1971 Feb. 19.

24. ESL. Library Board. *Annual Report.* 1931.

25. ESL. Library Board. *Minutes.* 1955 Sept. 15.

26. ESL. Library Board. *Annual Report.* 1961/62.

27. *Personal communication from Ralph Phelps.* 1979.

28. ESL. Library Board. *Annual Report.* 1979/80.

29. ESL. Library Board. Executive Committee. *Minutes.* 1915 Mar. 3, May 5.

30. ESL. Library Board. *Minutes.* 1921 Jan. 13.

31. Engineering Societies Library. *The reference collection of the Engineering Societies Library.* New York: the Library; 1970; 1970. 74 p. (ESL Bibliography; 17)

32. *A catalog of periodical publications in the library of the American Institute of Electrical Engineers.* New York: American Institute of Electrical Engineers; 1904.

33. *A catalog of periodical publications in the library of the American Institute of Mining Engineers.* New York: American Institute of Mining Engineers; 1904.

34. *Catalogue of technical periodicals: libraries in the City of New York and vicinity.* Compiled and edited by Alice Jane Gates. New York: United Engineering Society; 1915.

35. ESL. Library Board. *Annual Report.* 1915.

36. ESL. Library Board. *Minutes.* 1916 May 11.

37. ESL. Library Board. *Minutes.* 1966 Sept. 15; 1967 Sept. 21, Nov. 27; 1968 Feb. 15.

38. Engineering Societies Library. *Periodicals currently received as of Dec. 31, 1979.* New York: the Library; 1980.

39. ESL. Library Board. *Annual Report.* 1954/55; 1960/61.

40. ESL. Library Board. *Annual Report.* 1978/79.

41. ESL. Library Board. *Minutes.* 1921 Jan. 13; 1922 May 11.

42. ESL. Library Board. *Minutes.* 1913 Feb. 6.

43. ESL. Library Board. *Annual Report.* 1948/49.

44. Phelps, Ralph H. Engineering. *Library Trends.* 15(4): 868-879; 1967 Apr. (p. 876).

45. ESL. Library Board. *Minutes.* 1962 Jan. 18, May 17.

46. Kline, S. J.; Cabeen, S. K. Motion picture films as research data. *American Documentation.* 20(4): 385-386; 1969 Oct.

47. ESL. Library Board. *Annual Report.* 1979/80.

Chapter 8

1. Carnegie Library of Pittsburgh. *Annual Report.* 1910.

2. Hunt, Charles Warren. The activities of the American Society of Civil Engineers during the past twenty-five years. ASCE. *Transactions.* 82:1595; 1917.

3. Bay, J. Christian. *The John Crerar Library, 1895-1944: an historical report prepared under the authority of the Board of Directors.* Chicago: John Crerar Library; 1945: Chap. 2.

4. Weaver, William D., ed. *Catalogue of the Wheeler gift of books, pamphlets and periodicals in the library of the American Institute of Electrical Engineers, with introduction, descriptive and critical notes by Brother Potamian.* New York: American Institute of Electrical Engineers; 1909. 2 vols.

    5. ESL. Library Board. *Annual Report.* 1916.
    6. ESL. Library Board. Executive Committee. *Minutes.* 1919 Jan. 2.
    7. ESL. Library Board. Executive Committee. *Minutes.* 1919 Feb. 5, Mar. 5, Apr. 2.
    8. ESL. Library Board. *Annual Report.* 1919.
    9. *Ibid.*
    10. ESL. Library Board. *Minutes.* 1919 May 8.
    11. Dewey, Harry. Some special aspects of the classified catalog. *In:* Tauber, Maurice F., ed. *The subject analysis of library materials.* New York: Columbia University, School of Library Service; 1953: p. 114-129.
    12. Taylor, Kanardy L. Subject catalogs *vs.* classified catalogs. *In:* Tauber, Maurice F., ed. *The subject analysis of library materials.* New York: Columbia University, School of Library Service; 1953: p. 100-113.
    13. ESL. Library Board. *Annual Report.* 1919-25.
    14. ESL. Library Board. Executive Committee. *Minutes.* 1923 Mar. 27, Sept. 20.
    15. Raymond, Mary M. *The Engineering Societies Library, New York, and its classified catalog.* New York: Engineering Societies Library; 1936. Prepared for 28th annual conference of Special Libraries Association, Montreal, June 16-19, 1936.
    16. *Interview with Marguerite Soroka.* 1978.
    17. ESL. Library Board. Executive Committee. *Minutes.* 1937 Jan. 13.
    18. ESL. Library Board. *Minutes.* 1955 Jan. 20, Feb. 10.
    19. ESL. Library Board. *Annual Report.* 1959/60; 1967/68, 1969/70.
    20. *Classed subject catalog, Engineering Societies Library.* Boston: G. K. Hall; 1963.
    21. *Bibliographic Guide to Technology.* Boston: G. K. Hall; 1975-

    22. *Personal communication from S. K. Cabeen.* 1980.
    23. Shaw, Ralph R. *Engineering books available in America prior to 1830.* New York: Columbia University, School of Library Service; 1931. Master's essay.
    24. ESL. Library Board. *Minutes.* 1971 Feb. 19.

Chapter 9

    1. American Society of Civil Engineers. *Constitution, by-laws and act of incorporation.* New York: the Society; 1867: p. 4-5.
    2. Annual report of the Council. Report of the Library Committee. ASME. *Transactions.* 28:25-26; 1906.

3. United Engineering Society. *Charter and by-laws adopted December 16, 1904.* New York: the Society; 1904: p. 16.
4. Reports of the Council and Committees. The Library Committee. ASME. *Transactions.* 29:620-622; 1907.
5. ESL. Library Board. *Annual Report.* 1913-1917.
6. ESL. Library Board. Executive Committee. *Minutes.* 1915 May 5, May 14, June 10.
7. Report of the Library Committee for the year 1916. AIME. *Trans-actions.* 56:xxviii; 1917.
8. ESL. Library Service Bureau. *Letter [unaddressed].* 1915 Sept. 1. Located at: Engineering Societies Library, New York.
9. ESL. Library Board. *Minutes.* 1914 Feb. 5, Oct. 1.
10. ESL. Library Board. Executive Committee. *Minutes.* 1915 May 6.
11. ESL. Library Board. Executive Committee. *Minutes.* 1914 June 8.
12. ESL. Library Board. *Annual Report.* 1918-1945/46.
13. *Interview with Ralph Phelps.* 1977.
14. ESL. Library Board. Executive Committee. *Minutes.* 1919 April 2.
15. ESL. Library Board. *Minutes.* 1919 May 8, Oct. 9; 1920 Oct. 14; 1921 Jan. 13.
16. ESL. Library Board. *Annual Report.* 1923.
17. ESL. Library Board. *Minutes.* 1927 Oct. 13.
18. ESL. Library Board. Executive Committee. *Minutes.* 1924 June 19.
19. All library books now to be circulated to members. *Mechanical Engineering.* 62(4): 349-350; 1940 Apr.
20. ESL. Library Board. *Annual Report.* 1928 to 1945/6.
21. ESL. Library Board. *Minutes.* 1937 May 13; 1939 Jan. 12; 1940 Jan. 11.
22. ESL. Library Board. *Minutes.* 1953 Jan. 15, Mar. 12, June 18.
23. ESL. Library Board. *Minutes.* 1958 June 19.
24. ESL. Library Board. *Minutes.* 1950 Apr. 20.
25. ESL. Library Board. *Minutes.* 1973 Feb. 16, May 18, June 22.
26. ESL. Library Board. *Annual Report.* 1946/47-1979/80.
27. *Interview with Carmela Carbone and Carol Tschudi.* 1980.
28. ESL. Library Board. *Minutes.* 1953 Sept. 17.
29. ESL. Library Board. *Annual Report.* 1979/80.
30. *Ibid.*
31. ESL. *Bibliography on filing, classification and indexing systems, and thesauri for engineering offices and librarians.* New York: the Library; 1948-1966;1948-1966. (ESL Bibliography; 15) ESL Bibliography 15.
32. ESL. Library Board. *Annual Report.* 1979/80.

Chapter 10

1. The Annual Meeting, January 21, 1885. ASCE. *Proceedings.* 11:21-23; 1885.

2. Council of the Scientific Alliance. Building Committee. *Report.* New York: [the Alliance]; 1898.

3. *A catalog of periodical publications in the library of the American Institute of Electrical Engineers.* New York: American Institute of Electrical Engineers; 1904.

4. *A catalog of periodical publications in the library of the American Institute of Mining Engineers.* New York: American Institute of Mining Engineers; 1904.

5. *Catalogue of technical periodicals: libraries in the City of New York and vicinity.* Compiled and edited by Alice Jane Gates. New York: United Engineering Society; 1915.

6. Library dinner of the American Society of Electrical Engineers. AIEE. *Transactions.* 21:117, 119; 1903.

7. ESL. Library Board. *Minutes.* 1947 Oct. 16.

8. ESL. Library Board. *Minutes.* 1948 Feb. 24.

9. ESL. Library Board. *Minutes.* 1950 Jan. 19, Mar. 16.

10. *Interview with Ralph Phelps.* 1977.

11. *Personal communication from S. K. Cabeen.* 1981.

12. ESL. Library Board. *Minutes.* 1947 May 8.

13. ESL. Library Board. *Annual Report.* 1936/37.

14. ESL. Library Board. *Minutes.* 1967 Jan. 19, Feb. 16.

15. ESL. Library Board. *Minutes.* 1968 Nov. 19; 1969 Jan. 16.

16. *New York State union list of serials.* New York: CCM Information Corporation; 1970. 2 vols.

17. Fetterman, John. Resource sharing in libraries—why? *In:* Kent, Allen, ed. *Resource sharing in libraries.* New York: Dekker; 1974: p. 26.

18. ESL. Library Board. *Minutes.* 1967 Feb. 16.

19. Cabeen, S. K. The Engineering Societies library and the New York State Interlibrary Loan program: one special library's experience in a network. *Science & Technology Libraries.* 1(2):23-25; 1980 Winter.

20. ESL. Library Board. *Annual Report.* 1979/80.

21. Craver, Harrison W. Unfilled business. *Library Journal.* 63(12): 481-485; 1938 June 15 (p. 483).

22. Cabeen. *Op. cit.*

23. *Interview with Carmela Carbone and Carol Tschudi.* 1980.

24. ESL. Library Board. *Minutes.* 1930 Oct. 9.

25. ESL. Library Board. Executive Committee. *Minutes.* 1932 Feb. 17.

26. ESL. Library Board. Executive Committee. *Minutes.* 1947 Apr. 15.

27. ESL. Library Board. *Annual Report.* 1979/80.

Chapter 11

1. Hannum, Joshua Eyre. History of the Engineering Index. *Engineering Index.* xi-xiv; 1929.

2. Cutter, Walter [sic] Parker. An international technical index. *Special Libraries.* 2(8): 83-86; 1911 Oct.

3. ESL. Library Board. *Annual Report.* 1914, 1915.

4. ESL. Library Board. *Minutes.* 1917 Oct. 18, Nov. 7.

5. ESL. Library Board. Executive Committee. *Minutes.* 1918 May 1.

6. ESL. Library Board. Executive Committee. *Minutes.* 1918 June 12, Oct. 2.

7. ESL. Library Board. *Annual Report.* 1918.

8. ESL. Library Board. Executive Committee. *Minutes.* 1919 Jan. 2.

9. ESL. Library Board. Executive Committee. *Minutes.* 1924 Nov. 20.

10. ESL. Library Board. *Annual Report.* 1924.

11. ESL. Library Board. *Minutes.* 1925 May 14.

12. ESL. Library Board. Executive Committee. *Minutes.* 1925 Mar. 19.

13. ESL. Library Board. *Minutes.* 1927 Jan. 13.

14. Phelps, Ralph H. Engineering. *Library Trends.* 15(4): 869; 1967 Apr.

15. ESL. Library Board. Executive Committee. *Minutes.* 1928 Aug. 9; 1929 Mar. 14, June 20.

16. Craver, Harrison W. [*Letter to Earl Whitehorne.*] 1929 Mar. 11. Copy located at: Engineering Societies Library, New York.

17. ESL. Library Board. *Annual Report.* 1929.

18. ESL. Library Board. *Annual Report.* 1930; 1933.

19. Shaw, Ralph R. *The Engineering Societies Library: consultants in literature to consultants in technology.* [Gary, IN]; [1937]. Typescript. Copy located at: Engineering Societies Library, New York.

20. ESL. Library Board. *Minutes.* 1933 Oct. 5.

21. Shank, Russell. *Physical science and engineering societies in the United States as publishers, 1939-1964.* New York: Columbia University, School of Library Service; 1966: p. 150. Dissertation.

22. ESL. Library Board. *Minutes.* 1934 Oct. 11; 1940 May 9.

23. *Personal communication from Ralph Phelps.* 1977.

24. ESL. Library Board. *Minutes.* 1964 June 18, Nov. 19; 1965 Feb. 18, Dec. 9.

25. ESL. Library Board. *Minutes.* 1964 Sept. 17, Oct. 29; 1965 Jan. 21.

26. ESL. Library Board. *Minutes.* 1965 Sept. 16.

27. ESL. Library Board. *Minutes.* 1969 May 22, Oct. 30, Dec. 11.

28. ESL. Library Board. *Minutes.* 1970 Nov. 20.

29. ESL. Library Board. *Minutes.* 1971 Jan. 15, Feb. 19, Apr. 16, Sept. 17: appendix, Oct. 22.

30. ESL. Library Board. *Minutes.* 1972 Apr. 21.

Index

201